Endorsements

Science Careers: Personal Accounts from the Experts is a wonderful resource for anyone contemplating a career in the sciences. Lawrence Flowers has done a wonderful job gathering information from a variety of top-notch professionals, offering students an insider's view of many interesting science professions. I highly recommend this useful guide and wish it was available to me when I was in school!

—Mary C. Friesz, PhD, RD, CDE, LDN
Nutrition and Wellness Consultant & Author of *Food, Fun n' Fitness: Designing Healthy Lifestyles for Our Children*

I am excited about this book I believe the personal stories of people in different scientific professions will be inspiring. They can encourage students to consider a science career by seeing how these individuals pursued the path to their profession. Thank you for what I believe will become a very useful guide for those considering a career in the sciences.

—Marilynn E. Doenges, RN, MA, CS
Psychiatric nurse & co-author of *Nursing Care Plans, Guidelines for Patient Care, Nurses' PocketGuide, Diagnoses, Rationales, and Interventions, Application of Nursing Process and Nursing Diagnoses*

Science Careers

Personal Accounts from the Experts

Edited by
Lawrence O. Flowers

Foreword by Robert E. Yager

The Scarecrow Press, Inc.
Lanham, Maryland, and Oxford
2003

SCARECROW PRESS, INC.

Published in the United States of America
by Scarecrow Press, Inc.
A Member of the Rowman & Littlefield Publishing Group
4501 Forbes Boulevard, Suite 200, Lanham, MD 20706
www.scarecrowpress.com

PO Box 317
Oxford
OX2 9RU, UK

British Library Cataloguing in Publication Information Available

Library of Congress Cataloging-in-Publication Data

Science careers : personal accounts from the experts / edited by
Lawrence O. Flowers ; foreword by Robert E. Yager.
p. cm.
Includes bibliographical references and index.
ISBN 0-8108-4736-1 (pbk. : alk. paper)
1. Science--Vocational guidance. I. Flowers, Lawrence O.
Q147.S28 2003
502.3--dc21
2003001290

∞™ The paper used in this publication meets the minimum requirements of American National Standard for Information Sciences—Permanence of Paper for Printed Library Materials, ANSI/NISO Z39.48-1992.
Manufactured in the United States of America.

CONTENTS

FOREWORD

Many studies reveal that in excess of ninety percent of the American public is scientifically illiterate (Bybee, 1997; Miller, 1996). Such literacy has been a goal for science education for nearly 40 years as proclaimed by the National Science Teachers Association (NSTA). Science literacy has been defined by NSTA to be producing citizens who can:

- Use concepts of science and of technology and ethical values in solving everyday problems and making responsible decisions in everyday life
- Engage in responsible personal and civic actions after weighing the possible consequences of alternative options
- Defend decisions and actions using rational arguments based on evidence
- Engage in science and technology for the excitement and the explanations they provide
- Display curiosity about and appreciation of the natural and human-made world
- Apply skepticism, careful methods, logical reasoning, and creativity in investigating the observable universe
- Locate, collect, analyze, and evaluate sources of scientific information and use these sources in solving problems, making decisions, and taking action
- Distinguish between scientific or technological evidence and personal opinion and between reliable and unreliable information
- Remain open to new evidence and the tentativeness of scientific or technological knowledge
- Weigh the benefits and burdens of scientific and technological development
- Recognize the strengths and limitations of science and technology for advancing human welfare
- Connect science and technology to other human endeavors (e.g., history, mathematics, and the arts)
- Consider the political, economic, moral, and ethical aspects of science and technology as they relate to personal and global issues
- Offers explanations of natural phenomena that may be tested for their validity

Lawrence Flowers has offered a wonderful book—most of which comes from voices and experience of practicing scientists. Their own actions illustrate the features of scientific literacy as elaborated above. Their motivation, fascination, and personal experiences are recorded to serve as another way of illustrating scientific literacy that is not dependent upon their own scientific contributions to our understanding of the natural universe. Flowers's book provides a needed resource in every classroom for approaching scientific literacy and a way to encourage more to similar careers for lifetimes.

Instead of working to develop such competencies in every student, most teachers and schools are anxious to develop a curriculum that is often discipline-based and concerned with identifying basic concepts that students should know. These concepts are generally taught directly from textbooks and curriculum frameworks usually with teacher lectures and fact-oriented assessments to see if students can recite what they have been told or what they have been assigned to read.

Few have pondered what science is and what is needed to attain scientific literacy. Sagan (National Research Council, 1998) has argued that every child starts out as a scientist but formal education causes most to lose interest and to lose the basic qualities needed to do science. Sagan's point was that every child is full of questions, full of curiosity. And, every student also attempts to explain the things around him or her in ways that satisfy their own curiosity. Unfortunately, these personally formulated explanations are often in conflict with the lessons and theories that represent accumulated knowledge in the various sciences. However, these personally constructed explanations are nearly always more powerful than the explanations offered by teachers and textbooks. Some of the experiences of the scientists who have contributed to this book illustrate well the basic ingredients of science itself. The book, by the many examples, may be able to do what few school courses have been able to do. Teachers may be able to draw from the experiences, perceptions, and motivation of current scientists in ways not possible in most classrooms. And, these exciting reports may also affect some of those most curious to similar careers.

Simpson (1963) has developed a meaningful definition of science that should be used in analyzing professional work and careers in the sciences. He outlines several features of the scientific enterprise, namely:

- Asking questions about the natural universe (i.e., being curious about the objects and events in nature)
- Trying to answer one's own questions (i.e., proposing possible explanations)
- Designing experiments to determine the validity of the explanation offered
- Collecting evidence from observations of nature, mathematics calculations, and, whenever possible, experiments carried out to establish the validity of the original explanations

Late in the 1960s, career education became an important undertaking in most schools; this was certainly true in science. Unfortunately, however, this focus often became brief biographical sketches of some of the famous scientists of the past. They were often dead and important in terms of firmly established theories and explanations of the natural world that they had offered. They always focused on the past, and seldom captured the essence of the science activities of the present. Flowers's book can help students and teachers with actual experiences of living scientists offering a new focus and a new and more realistic way of viewing the scientific enterprise in school.

In the 70s, Harms received a large grant from the National Science Foundation (NSF) that was written with the identification of four goal clusters that defined the desired domains for school science. These goals (Harms, 1977) included:

- Science for meeting personal needs. Science education should prepare individuals to use science for improving their own lives and for coping with an increasingly technological world
- Science for resolving current societal issues. Science education should produce informed citizens prepared to deal responsibly with science-related societal issues
- Science for assisting with career choices. Science education should give all students an awareness of the nature and scope of a wide variety of science and technology-related careers open to students of varying aptitudes and interests
- Science for preparing for further study. Science education should allow students who are likely to pursue science academically as well as professionally to acquire the academic knowledge appropriate for their needs

Obviously, the third goal was a new look at a focus on career awareness for all students concerning the fascinating life and profession for scientists. The focus on goals for school science was considered anew in the National Science Education Standards (NRC, 1996). This $7 million effort over four years was organized once more with four goals. These include:

- Experience the richness and excitement of knowing about and understanding the natural world
- Use appropriate scientific processes and principles in making decisions
- Engage intelligently in public discourse and debate about matters of scientific and technological concern
- Increase their economic productivity through the use of the knowledge, understanding, and skills of the scientifically literate person in their careers

School and college science tends to ignore these basic goals. Instruction too often is but a review of what good scientists agree are our best interpretations of nature today. They do not concentrate on any connection to current living, to preparing students to assume citizenship responsibilities, or to consider concerns and actions designed to improve economic productivity. Practically no attention is given to encouraging every student to have even one experience with every aspect of science—like the features of the enterprise involved in Simpson's definition. Flowers's book suggests the importance of the National Standards, of the importance of viewing science in its broadest sense, of portraying science as a kind of thinking and analysis that is personal and unique for each individual.

Science is basic to every human. And yet is rarely seen by students as solving mysteries of the natural world or of getting them so engaged that it becomes a career. Perone (1994) has compiled what research says about getting minds engaged. His list includes:

- Students must help define the content often by asking questions
- Students must be given time to wonder and to find interesting pursuits
- Teachers encourage and request different views and forms of expression
- Students create original and public products that enable them to be "experts"
- Students take some action as a result of their study and their learning
- Students sense that the results of their work are not predetermined or fully predictable

Perhaps the greatest gain for readers of Flowers's book is to see science as something people do. Capturing their fascination, their unique contributions, and the give and take among professionals all represent the lessons this book can provide. Science basically is producing evidence of the validity of human conceptions of how the natural world operates. Science is not done in a vacuum. The scientific ideas must be accompanied with observations and experimental evidence that others can see. The community of scientists must accept the explanations as valid. This is unlike other human endeavors such as music, art, literature, religion, and the humanities. One would never tamper with a masterpiece of art or literature. In science, however, every explanation is subject to scrutiny, to tampering, to considering degrees of accuracy. Scientific explanations are tentative. They exist only to help with further and better explanations. This guidebook, with statements and stories by professionals, is meant to help with the issue of scientific literacy while also encouraging students and others to grow in terms of this literacy and to make better judgments regarding careers.

Curious people and creative people who like the mysteries of the natural world can have satisfying careers in real science. Hopefully, reading about the lives of motivated, curious, creative scientists who are alive and active today will attract the kind of new people in science that we need. Perhaps these people can all help in our efforts to attain a truly scientifically literate society. This is the potential of this book. This potential indicates the importance of this book for attracting the needed persons to be future scientists. Persons with the needed attributes and motivation are needed to assure a sustainable future and a reversal of the degradation of the earth. Never has there been a more critical time for more and better people in scientific careers—and helping prepare a citizenry that is scientifically literate and able to assist in making societal decisions. Flowers and the contributing scientists have contributed more by making this book available for teachers, parents, and students.

Robert E. Yager, Ph.D.
Professor of Science Education
University of Iowa

Dr. Yager is a science education professor in the department of science education at the University of Iowa. Dr. Yager has been president of seven national organizations: National Science Teachers Association, National Association for Research in Science Teaching, National Association of Biology Teachers, Association for the Education of Teachers in Science, School Science and Mathematics Association, American Association for the Advancement of Science, and the National Association for Science-Technology-Society. Furthermore, Dr. Yager has directed over one hundred National Science Foundation funded projects and has published over five hundred scholarly publications.

References

Bybee, R. *Achieving scientific literacy: From Purposes to Practices*. Portsmouth, NH: Heinemann, 1997.

Harms, N. C., & Yager, R. E., eds. *What Research Says to the Science Teacher*, Vol. 3. Washington, DC: National Science Teachers Association, 1981.

Miller, J. D. *Scientific Literacy for Effective Citizenship*. Edited by R. Yager, *Science/Technology/Society as Reform in Science Education.* Albany, NY: State University of New York Press, 1996.

National Research Council. *National Science Education Standards*. Washington, DC: National Academy Press, 1996.

National Research Council. *Every Child a Scientist: Achieving Scientific Literacy for All*. Washington, DC: National Academy Press, 1998.

Simpson, G. G. "Biology and the Nature of Science." Science 139 (1963): 81– 88.

PREFACE

The purpose of this book is to assist high school, undergraduate, and graduate students in preparing for a career in the sciences. The valuable information presented in this book will also aid academic counselors, advisors, and career service counselors who are likely to advise students about possible careers in the sciences.

When I initially considered writing this book, I had just completed my fourth year of teaching undergraduate science students. As a result of various discussions with numerous students regarding science careers, I became increasingly aware of the importance of a book that would provide high school, undergraduate, and graduate students assistance in critically understanding relevant information regarding specific careers in the sciences. Therefore, the origin of this guidebook stems from countless experiences with students regarding science career explorations. The ever-increasing body of published literature and an increasing number of science careers have made it necessary for this text. If you are thinking about a career in the sciences or if you are conducting research in order to narrow down your career search, this book should be of great value to you.

The information presented herein reflects a diverse set of experiences from scholars, professionals, and practitioners actually working in science-related fields. The recommendations contained in this work will introduce the reader to pertinent discourse regarding careers in the sciences; however, it is not designed to address all the concerns of every student.

The current manuscript aims to: 1) engage students in active discourse regarding careers in the sciences, 2) help students develop intelligent, well-informed views regarding careers in the sciences, 3) use a systematic pedagogical approach to dealing with science careers, and 4) help students be aware of the hazards and pitfalls associated with selected careers in the sciences. Moreover, this book contains a collection of straightforward, biographical sketches of dedicated scholars, professionals, and practitioners in science-related careers all designed to inspire, motivate, and guide students as they consider a career in the sciences. Specifically, the information in this book is designed to provide students with a general overview of a particular science career.

The present career exploration resource guide goes beyond simply offering a superficial description of select science careers. For this guide science professionals were recruited to submit an original contribution regarding their particular career to take advantage of their experiences. The result is a guide book that presents an in-depth representation of selected science careers by the men and women currently participating in these exciting fields. It is the hope that this book may act as a stimulus for further discussions and interviews with science professionals currently working in your intended science career.

The current manuscript achieves a new distinction. No other career exploratory resource guide currently in print can profess to address the issue of science careers in the manner attained in this helpful guide. This work is novel in several ways. First, each contribution is divided into thirteen components (described in more detail in chapter 1). The organizational scheme is designed to explore selected science careers adequately and in a way that will be highly accessible to students. In addition, this plan has made it possible to give comprehensive coverage to the important factors regarding career exploration.

This book is organized into four chapters. Chapter 1 will introduce the reader to the scope and organization of the book. Chapter 2 will present interesting information regarding science careers compiled from the Bureau of Labor Statistics. Chapter 3 contains personal accounts from various science professionals. Chapter 4 is devoted to providing students with additional resources such as reliable science career websites and key contact information from science professional associations and societies.

The innumerable insights contained in this manuscript are the result of the efforts and experiences of the science professionals featured in this book. These individuals collectively have amassed over two hundred years of experience of which students can take full advantage. This book also offers essential explanations on how to process and assimilate the useful information contained in this resource guide.

So rapidly does knowledge expand that there is an enormous need for resources that can make order out of chaos, sense out of nonsense, and bring calm when there once was confusion. This manual purports to fulfill these requirements and more. The editor appreciates the fact that much of what students learn in school is soon antiquated and fails to teach students adequately about possible careers in a particular area. The study of science careers can make a permanent and important contribution to the overall aptitude of the student.

It is the ultimate hope of the editor that this book will serve as the first of a series of many books regarding specific science careers. The proposed series, while retaining the pedagogical framework found throughout this book, will address individual scientific disciplines.

The overriding impetus behind the production of this text reflects a sincere desire to help students understand and appreciate the complexity of decision-making processes associated with career exploration. Moreover, the overall aim of this book is to educate science students with knowledge pertaining to the necessary skills relevant to select science careers. Further, the mission of this text is to instill in students the aspiration for continual improvement to ensure a successful career and to provide a text filled with supportive and understandable language designed to benefit all science students in a deep and profound way.

Furthermore, the views and opinions expressed in this manuscript are those of the editor and contributors. I earnestly welcome any comments and suggestions from students and science professionals alike for future and subsequent revisions of this resource.

Keen attention has been given to create a text that applies essential pedagogical approaches throughout the manuscript, thereby offering students the necessary support needed to effectively apply the topics explored in this book and to derive maximum benefit from this book.

In order to make the book more visually appealing, a clear, crisp, and large style font has been employed throughout this text. The editor feels that the organization of the manual provides readers with a great deal of satisfaction that comes from gaining critical knowledge today that will undoubtedly secure a more productive and rewarding tomorrow. It is my sincere hope that the stories, words, advice, ideas, and insights in this book will be helpful as you consider a fascinating career in the sciences.

Lawrence O. Flowers
Book Editor

ACKNOWLEDGMENTS

The current text represents the efforts, endeavors, and expertise of many individuals. It is an honor and a privilege for me to acknowledge those responsible for *Science Careers: Personal Accounts from the Experts*.

I am particularly grateful to the science professionals who have provided immeasurable insights, experiences, and knowledge regarding their particular science career.

I express warm thanks to Kim Tabor, my acquisitions editor, for her constant support, helpful advice, understanding, enthusiasm, and patience. Her perseverance contributed to the high quality of this text. I am also grateful to all of the supporting staff at Scarecrow Press for their meticulous attention to detail.

Additionally, I would like to thank all of the many persons that have offered fruitful suggestions, constructive criticisms, novel ideas, and editorial assistance toward the improvement of this book. Although all of these valuable recommendations could not be employed, many have been faithfully incorporated in the book. These generous communications played a major role in the shaping and development of this manuscript.

I especially acknowledge Dr. Robert E. Yager for his willingness to provide the foreword for this manuscript. The superbly crafted information contained in the foreword made a significant educational contribution to the book.

A special note of thanks is extended to my family and friends for their encouragement and unconditional support during the writing, editing, and processing of the book. To my parents, Alvin and Gloria Flowers, thank you for instilling in me the positive values necessary to be successful in every aspect of life.

This book is also dedicated to the memory of very special friends and family who are no longer with us. These extraordinary individuals found the time to teach and share in a unique manner and have greatly shaped my life in a profound and meaningful way.

The editor assumes full responsibility for any errors, omissions, and inaccuracies that may appear in this text. Moreover, the editor feels that the current text must be responsive to the needs, criticisms, and opinions of students, professors, counselors, and career service professionals and therefore would appreciate any comments and suggestions for improvement of this resource guide. A special gratitude is owed to those students who have asked for clarification of certain aspects of career decisions and who have continued to seek further information in regards to careers in the sciences.

Furthermore, I am deeply indebted to many nameless mentors, advisors, teachers, professors, counselors, colleagues, and special friends. I am also grateful to numerous instrumental persons from Harold M. Ratcliffe Elementary School, George H. Moody Middle School, Henrico High School, Virginia Commonwealth University, University of Iowa, and the University of Florida who have guided my professional development and therefore made it possible for me to complete this book.

CONTRIBUTORS

The following persons listed below have offered key contributions to this book. A sincere appreciation is directed to these dedicated persons who are deeply committed to the educational improvement of all students.

Dr. Lamont A. Flowers
Assistant Professor
Department of Educational Leadership, Policy, and Foundations
University of Florida

Dr. Peter H. Gilligan
Clinical Microbiologist
Clinical Microbiology-Immunology Laboratories and Phlebotomy Services
University of North Carolina Hospitals

Mrs. Barbara W. Harrison
Certified Genetic Counselor
Department of Pediatrics and Child Health
Howard University

Dr. Sherman S. Hom
West Nile Virus & Bioterrorism Lab Coordinator
Division of Public Health & Environmental Laboratories
New Jersey Department of Health & Senior Services

Mr. Toby Q. Jenkins
Industrial Engineer
Liquids Organizational Effectiveness Manager
Proctor & Gamble Manufacturing Company

Mrs. Natasha J. Johnson
Registered Pharmacist
Pharmacy Department
Eckerd Drug Company

Mrs. Karen M. Kiser
Professor
Clinical Laboratory Technology and Phlebotomy
St. Louis Community College at Forest Park

Dr. James L. Moore III
Assistant Professor
School of Physical Activity and Educational Services
Ohio State University

Dr. Tonya L. Peeples
Associate Professor
Department of Chemical and Biochemical Engineering
University of Iowa

Dr. Karen Joy Shaw
Team Leader, Infectious Diseases
Research Fellow
Johnson & Johnson Pharmaceutical Research & Development, L.L.C.

Mrs. Valerie J. Shereck
Adult Nurse Practitioner
Penrose-St. Francis Senior Health Center
Centura Health

Dr. Amy L. Springer
Biotechnologist
Senior Scientist
Prolinx, Inc.

Dr. Mark F. Vondracek
Physics Teacher and Research Advisor
Science Department
Evanston Township High School

Dr. Carol J. Zimmerman
Geophysical Associate
Geophysical Applications Group
ExxonMobil Exploration Company

CHAPTER 1

Introduction

The process of selecting a career in the sciences is a complex endeavor. It involves many factors from selecting a subject you enjoy to preparing for your initial interview. While the activities mentioned above are major elements in terms of selecting a career and ultimately getting hired, this book will only focus on a specific aspect of career exploration. Specifically, the current text will:

- Examine issues important to future science professionals
- Present insightful views by current science professionals
- Include tips on how to avoid science career-related pitfalls
- Contain information on science professional associations
- Offer information on how to prepare for a science career

Unlike most books written about science careers that merely present a detailed listing of job descriptions, this book offers credible discourse on particular careers in the sciences by the men and women who participate in these exciting careers. Additionally, the purpose of this book is to:

- Introduce readers to actual science professionals
- Assist students in the career decision-making process
- Help students think about careers in the sciences
- Encourage students to intelligently discuss science careers
- Communicate high expectations regarding science careers

What Is a Science Career?

A general book on science careers, such as the case for this book, should contain a broad and diverse selection of topics. This book contains such a selection and will be of interest to every science student, no matter the education level. However, it begs the question, what is a science career? For the purpose of this book, a science career is defined as a career based on a systematized knowledge encompassing general truths. Alternatively, a science career is a career based on the operation of general laws regarding the physical world and its inhabitants obtained and tested through experimentation. Based on the previous definitions, this guide will include contributions from the fields of engineering, biology, microbiology, geology, physics, pharmacy, biotechnology, genetics, and virology.

It should be noted that science careers, like most other careers, take place in many different settings. For example, science professionals can work in industry, business, hospitals, private sector, academia, and federal or local governmental settings. Within each setting there are a wide variety of positions available to science professionals. Some of these positions will be discussed in this book. The brief list that follows is a very small sampling of the types of jobs available in the sciences: medical doctor, administrator, laboratory assistant, professor, laboratory technician, laboratory director, research assistant, clinician, nurse, veterinarian, pharmacist, dentist, genetics counselor, chemist, engineer, medical technologist, pathologist, dietitian, and science school teacher. This is in no way an exhaustive list of potential science careers. You are advised to consult additional specialized texts and articles to learn more about the many different positions available in science-related fields.

What Is an Expert?

The term *expert*, as used in this text, will be broadly defined as an individual who has a special skill or knowledge representing mastery of a particular subject area or job duties based on job tenure. The experts in this manual have been very successful in their current professions and have amassed a considerable amount of insights and advice designed to increase your awareness of select science careers. It is estimated that the experts in this manual have a combined work experience of over two hundred years from which you can benefit.

Organization and Content

This book begins with a general discussion on the scope of the book and then examines specific careers by addressing issues such as, advantages, pitfalls to avoid, and pertinent skills needed to ensure a successful career. The primary objective of this book, as well as similar qualitative literary works, is to obtain key information by examining science careers by questioning current science professionals working in a particular science career. Given the nature and scope of this book, the qualitative component in this book is necessary and advantageous to thoroughly explore different science careers (Lincoln & Guba, 1985). The approach employed in this book will allow for a more in-depth look into the opinions and experiences of science professionals. This book is divided into four chapters:

Chapter 1: Introduction

This chapter introduces the reader to the text and provides an overview of the content of the book including the purpose of the book and intended audience. This section also includes sample career networking letters and explains a very practical conceptual model that will enable prospective science professionals to make the correct and most informed decision regarding their science career.

Chapter 2: Science Careers: Statistical Exploration

Chapter 2 provides interesting and useful statistical information about science careers. This chapter presents information on the current numbers of science graduates, salary information on specific careers, and useful employment data pertinent to those interested in a career in the sciences.

Chapter 3: Personal Accounts from the Experts

This chapter represents the main thrust of the book. Here, science professionals from science careers provide personal accounts of their career. Each contribution is arranged in the same format:

Name of contributor
Presents the full name of the contributor

Job title
Presents the job title of the contributor

Institutional affiliation
Provides the name of the contributor's occupational institution

Favorite quote
Presents contributor's favorite quote regarding goals, science, etc

Biographical Sketch
Presents biographical information about the contributor

Description of Job Duties
Provides an honest and candid description of a particular science career

Advantages of Career
Presents the benefits of working in a particular science career

Essential Related Skills for Success
Discusses what additional skills are needed for a successful science career

Additional Advice
Offers further recommendations and suggestions for a successful science career

Pitfalls to Avoid
Presents potential obstacles and solutions regarding a particular science career

Qualifications Required
States educational and professional requirements needed for science career

Take Home Points
Summarizes main points of contribution for quick review

Suggested Reading List
Presents pertinent resources for further career exploration

Chapter 4: Science Careers: Additional Resources

This chapter will feature additional helpful information such as the latest and most useful websites devoted to science careers arranged in a well-organized format. Chapter 4 also contains a Career Evaluation Form, a list of professional associations categorized by science discipline, a Science Career Network Investigation Form, and science career network questions. These resources are designed to assist you with your career search and help you make the best decision regarding your professional future.

The Decision Cycle

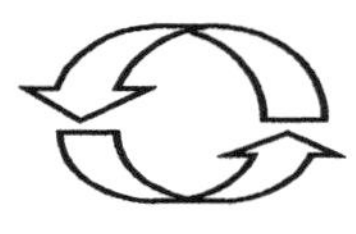

In the process of exploring different career options one must collect as much information about the particular career as possible. This arduous and essential activity can take many forms such as, conducting interviews with science professionals, utilizing the Internet, reading about science careers, or actually performing the intended job duties to get an idea of the type of work associated with a given career. The authors of *Discover the Career Within You* have created a useful model you can employ to help make an informed career decision (Carney & Wells, 1995, pp. 20–24). Their model, "The Decision Cycle," emphasizes seven distinct stages involved with making an intelligent career decision: awareness, self-assessment, exploration, integration, commitment, implementation, and reevaluation. The text that follows provides a summary of the stages associated with "The Decision Cycle."

1. Awareness
This first stage involves the recognition that a career decision needs to be made. The onset of this stage can occur for many reasons such as an upcoming graduation, selection of an appropriate major, or the realization that you must pursue a career change. Whatever the reasons may be, it is at this stage that you become conscious that you must make a decision regarding your future. The fact that you are reading this book to learn information about specific science careers suggests that you may have already embarked on this stage.

2. Self-Assessment
This stage involves learning about your own personality, interests, desires, strengths, and weaknesses in order to choose a career that maximizes your talents and minimizes your deficiencies. This information can easily be obtained with the use of standard career assessment surveys and questionnaires as well as widely accepted personality indicators such as the Myers-Briggs Type Indicator, Holland's Self-Direct Search, and the Strong Interest Inventory (Myers & McCaulley, 1985; Holland, Fritzsche, & Powell, 1994; Harmon, Hansen, Borgen, & Hammer, 1994).

3. Exploration
This stage involves the acquisition of specific knowledge, relevant skills, or experiences pertaining to different and varied science careers. It includes obtaining essential information about your intended career by reading specialized career books, perusing magazines and newspapers, networking, conducting interviews, participating in internships and summer programs, exploring the Internet, participating in professional associations, attending career fairs and informative seminars, engaging in job shadow or volunteer endeavors, and subscribing to trade magazines, to name several approaches.

This type of information acquisition must involve a detailed and systematic exposure to a number of different science careers. The exploration stage, like the other stages in this cycle, is complex and can encompass interactions between many persons and requires a great deal of dedication and persistence on your part.

4. Integration
The fourth stage, integration, involves a detailed and meticulous analysis of the information you uncovered in stages 2 and 3. In this stage you utilize all available information by research and experiential endeavors to select the best and most beneficial career choice.

To perform this type of analysis, employ the Career Evaluation Form. The Career Evaluation Form is a document that allows you to visually examine the advantages and disadvantages of each prospective science career. The left side of the page is labeled Advantages of Career. The right side of the page is labeled Disadvantages of Career. On the bottom of the page is a space for you to write your most cherished educational and personal lifelong goal(s). Fill out the Career Evaluation Form for each of your prospective science careers. Figure 1.1 displays the Career Evaluation Form. The most favorable science career choice is the choice in which the advantages outweigh the disadvantages and most aptly meets your educational and personal goals and matches your interests, desires, and skills.

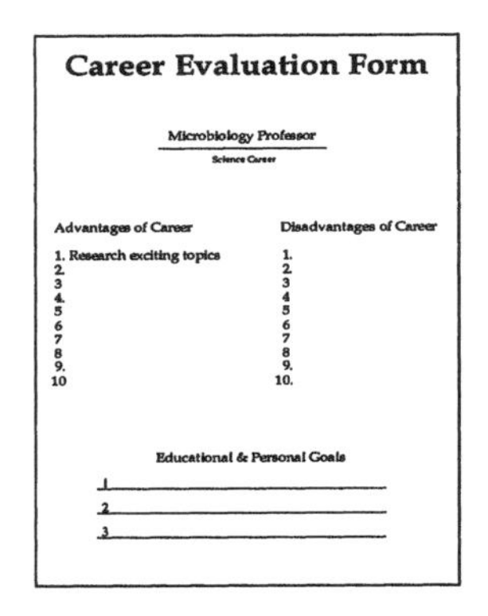

Career Evaluation Form

Microbiology Professor

Science Career

Advantages of Career	Disadvantages of Career
1. Research exciting topics	1.
2.	2.
3	3
4.	4
5	5
6	6
7	7
8	8
9.	9.
10	10.

Educational & Personal Goals

1 ______
2 ______
3 ______

Figure 1.1 Career Evaluation Form

5. Commitment

Based on the evaluation performed in the previous stage you are now ready to make an informed career decision. During this stage you must commit to do whatever it takes to pursue your career choice no matter the costs, risks, or obstacles you may face along the way.

6. Implementation

After you have made an informed decision regarding your future, the next step is to devise and document a plan for actually achieving your career goals. This stage involves:

- Creating a calendar highlighting short-term goals
- Making contacts with potential mentors and employers
- Acquiring new skills and honing old skills
- Obtaining additional training in deficient areas
- Taking specific courses
- Graduating
- Gaining work experience
- Volunteering

As you know, merely documenting a plan is meaningless if you do not stick to it. Remember, persistence leads to success.

7. Reevaluation

The final stage in this cycle involves the reevaluation of your career decision. During the implementation stage you must constantly assess whether the decision you have made is in accordance with your educational and personal goals while satisfying your self-interests. Change is inevitable; it is important to remain open-minded to the possibility that you may have to alter your career decisions to meet the current demands of your life. Reevaluation also involves minor adjustments in the plan you created in the previous stage. Lastly, re-evaluation can lead to the realization that your first career choice is no longer the best choice and that it may be more advantageous to start the decision cycle over again.

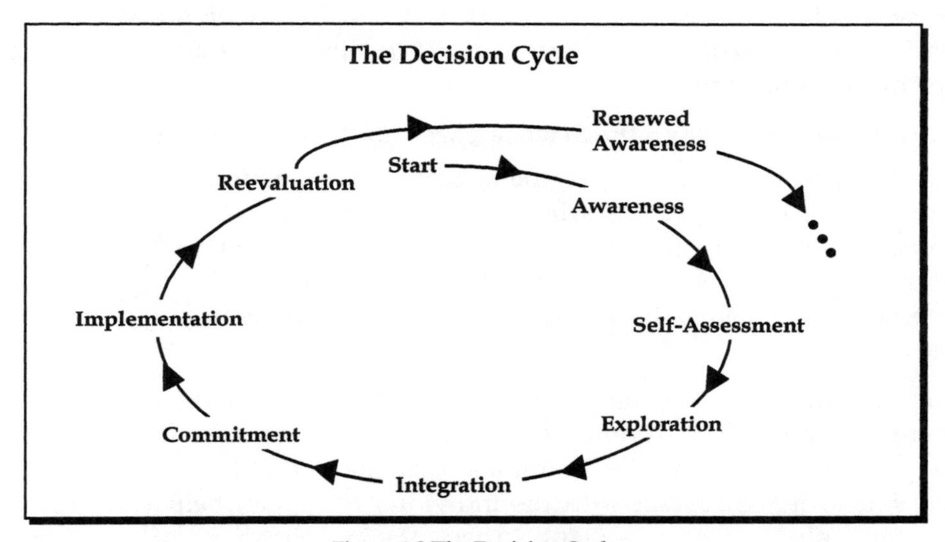

Figure 1.2 The Decision Cycle

Source: From *Discover the Career Within You*, 4th edition, by C. G. Carney and C. F. Wells © 1995. Reprinted with permission of Brooks/Cole, an imprint of the Wadsworth Group, a division of Thomsom Learning. Fax (800) 730-2215.

The decision cycle is shown in figure 1.2. Each stage in the decision cycle is comprised of separate, yet complementary, variables that interact to shape one's overall career decision. Specifically, this book is designed to promote the exploration stage, the stage in which the prospective employee carefully investigates various aspects of an intended career. This investigation leads the future employee to evaluate different career options and then to a decision regarding the best career option based on valuable information. The main practical tenet regarding the decision cycle can be summarized in two sentences: Before you actually pursue a science career learn all that you can about yourself and the intended career by talking to future employers and gathering useful personal and career-related information. Once this information has been obtained consider the pros and cons regarding the career, make an informed decision, and then go after it.

There are usually instrumental persons on your high school or college campus or in your local community trained to help you make the best career decision for you. Those helpful individuals include career counselors, professors, teachers, advisors, mentors, and friends to name just a few. Your career decision is probably one of the most important decisions you will ever make. Please make sure this paramount decision is not made out of haste or lack of key information. In total, the career exploration process must involve the collection of useful data and information about a wide variety of different career aims in order to make an informed decision regarding your future.

Career Networking and the Science Career Network Investigation Form

As stated in the preceding segment, career investigation can involve many persons who may assist you in your career choice. From the awareness stage to the reevaluation stage, a support system in which information and special services are shared among essential people is paramount.

Networking is an essential component in the career exploration process. Networking involves forming relationships with people for the purpose of helping you succeed in life. You must understand that it is very important for you to make contact with the individuals that can help you grow in educationally appropriate ways. In fact, the establishment of an effective career network can lead to internship opportunities, valuable career advice, research opportunities, job shadow prospects, and more.

Science Career Network Investigation Form

Prospective Science Career: *Quality Control Supervisor*

Contact Information

Name: Dr. Lauren Mills

Address: 1985 Sunny Lane, Flower Garden, TX 02145

Telephone Number: (123) 987-6543 **Fax Number:** (197) 405-0534

Job Title of Contact, Institutional Affiliation: V.P., Genetics Corp.

E-mail Address: lmills@genecorp.com

Additional Notes: Contact Dr. Mills to discuss internship opportunity.

Figure 1.3 Science Career Network Investigation Form

Just as the detective investigates to find more clues about a particular crime you must also thoroughly investigate every opportunity to help you choose the perfect science career for you.

The Science Career Network Investigation Form is designed to allow you to record essential contact information about members of your career network. Use the information in this form to keep track of the type of service or support that members of your career network provide. Figure 1.3 displays the Science Career Network Investigation Form. A blank form is found in chapter 4 for your personal use.

Career Networking Letters

Below are samples of key networking letters that will undoubtedly assist you as you make contact with members of your career network. A timely placed letter can certainly promote a beneficial working relationship and can also increase your opportunities to learn more about your prospective career. The four letters provided herein (see figs. 1.4–1.7) should be used as you delve into the career investigation process. While all of the letters found in this section are essential for your success, the most important letter is the thank-you letter. In my experiences, I have found that students fail to utilize this letter effectively. This point can't be overemphasized. If you are fortunate enough to make contact with a person that selflessly offers their time, advice, counsel, or services it is your responsibility to thank them in the form of a thank-you letter.

June 7, 2004

Mrs. Barbara Lavier
Biology Corporation
0607 Garner Avenue
Richmond, Virginia 24680

Dear Mrs. Lavier:

Thank you for your advice regarding my career decision. I am extremely grateful to you and appreciate the time and effort you spent on my behalf. It was indeed a pleasure meeting you.

Thank you again for your assistance.

Cordially,

Nora Velanis

Nora Velanis

Figure 1.4 Sample Thank-You Letter

February 12, 2004

Dr. Royal Jones
Professor, Cierra University
2601 University Avenue
Amelia, Georgia 48024

Dear Dr. Jones:

My name is Terry Evans. I am a junior majoring in microbiology at Cierra University. Last semester I took your course entitled, *Introduction to Microbiology*. I am writing to setup a meeting with you at your earliest convenience regarding my desire to work in your laboratory. Please let me know what time is most convenient to meet.

Thank you in advance for your time and consideration.

Sincerely,
Terry Evans
Terry Evans

Figure 1.5 Sample Networking Letter

January 15, 2004

Dr. Richard Elder
Science Professionals Society
5555 Career Street
Iowa City, Iowa 12345

Dear Dr. Elder:

My name is Brenda Pittman. I am a sophomore at Cummings University majoring in genetics. I would be pleased if you would send me membership information regarding the Science Professionals Society.

Thank you for your time.

Sincerely,
Brenda Pittman
Brenda Pittman

Figure 1.6 Sample Letter Requesting Materials for Membership of Professional Association

May 12, 2004

Mr. Raymond Brown
Science Lab Industries
6792 Biology Lane
Russell, Florida 35791

Dear Mr. Brown:

It was a pleasure to meet with you last week at the Science Career Fair at Harris College on May 5, 2004. As I informed you during our brief meeting I am very interested in learning about the research environment in your company's Biotechnology Division. Please contact me at (515) 370-1223 at your earliest convenience.

I look forward to having the opportunity to meet with you again to discuss any possible internship endeavors. Thank you very much for your time.

Sincerely,

Bruce Munroe

Bruce Munroe
Graduate Student, Harris College

Figure 1.7 Sample Follow Up Letter

How to Use This Book

To better assist you in the use of this book I have provided a few tips that will not only guide you as you read but will allow you to get the most from the text. These tips include:

1. As you learn more about your intended science career, as a result of reading this book, create useful study aids by utilizing the Career Evaluation Form to aid in the career selection process.

2. As you read the *Description of Job Duties* and *Essential Related Skills for Success* sections for your science career of interest, introspectively examine your own personal or academic strengths and weaknesses. Pontificate and implement ways in which to improve or strengthen various areas to ensure a successful science career.

3. The information contained in the sample letters can also be employed when making telephone calls. When career networking using the telephone: be brief, state the precise nature of your call, and be extremely courteous and polite.

4. Record interesting facts and details about your prospective science career as

you engage in the career exploration process. Chapter 4 contains a personal journal for you to conduct this rewarding activity. This information can then be transferred to your Career Evaluation Form.

5. At the end of each contribution is a suggested reading list. This list contains additional books about a particular science career. If more explanation is required on a certain topic please use these valuable sources as a starting point for further clarification.

6. Chapter 4 contains useful websites that are of special interest to those interested in pursuing a science career. This resource is organized into specific scientific areas. Explore these informative websites to learn more about a particular science career, post your resume, discover employment opportunities, and more.

7. Chapter 4 also contains a list of contact information for various science-related professional associations. Write to the organization of your choice to learn more about your intended science career. Also attend meetings and seminars presented by the professional organizations of your choice. Use this opportunity to expand your career network and also to talk to other students who are interested in similar career issues.

8. The blank forms located in chapter 4 (e.g., Career Evaluation Form and the Science Career Network Investigation Form) are for your personal use. Feel free to photocopy these forms if needed.

Due to practical limitations only a small number of science careers were selected for inclusion in this book; therefore, your intended science career may not be represented here. Don't worry—in fact the most important and valuable aspect of this book is the actual format of each contribution. The format of each contribution should serve as the underlying basis for all of your career investigation endeavors. In other words, the contribution format used in this book should serve as your initial inquiries in your pursuit of acquiring knowledge about your intended science career. Additionally, each contribution (regardless of the science career) in this book contains valuable insights and helpful information you can use to prepare for your career. I suggest that you read each contribution even if the career discussed does not directly relate to your current goals. In the near future I hope to bring you another installment of this series that addresses your specific career.

Concluding Remarks

The purpose of *Science Careers: Personal Accounts from the Experts* is to serve as a helpful tool to assist science students and prospective science professionals in learning about select careers in the sciences. Moreover, this book contains a collection of straightforward, biographical sketches of current scholars, professionals, and practitioners in science-related careers. As a result of reading this book you should feel like you have just conducted your own personal interview with each contributor. The personal accounts should enable you to make clear connections between classroom instruction and real-world applications.

Overall, the intended goal of this book is to reduce your current misunderstandings regarding select science careers, increase your awareness of various pitfalls associated with select science careers, and make the career investigation process a rewarding experience. Furthermore, this book will greatly add to the career exploration literature currently available by offering an innovative look at potential careers in the sciences. Additionally, this book seeks to address the educational needs of students and assist secondary education and university career counseling professionals in providing straightforward accounts of science careers from persons that currently work in the field.

As previously mentioned the intended audience for the book includes a number of constituent groups. First, the general audiences for this book are undergraduate and graduate science students interested in pursuing employment in science-related fields. The information in this book is equally beneficial to high school students interested in majoring in science-related fields and ultimately pursuing a science career. Thirdly, since this book will contain personal accounts from scholars, practitioners, and professionals in science-related fields it will serve as a credible source for academic advisors and career service counselors.

Your science career, like most major decisions, must be chosen for the right reasons. Specifically, your decision to enter a particular science career must be based on well-researched information. Your decision must also focus on your interests, academic strengths, and personal traits. Simply wanting to enter into a science career because of family history is not an adequate reason. While there are many advantages for choosing a career in the sciences, if you make the wrong decision based on haphazard career investigation techniques you will have the rest of your life to regret your decision. It is my hope that the stories, experiences, observations, and recommendations presented in this book will motivate, inspire, and educate all students interested in pursuing a science career. Finally, as you begin to embark on your career exploration journey I leave you with a final thought from Confucius:

"Choose a job you love, and you will never work a day in your life."

Take Home Points

- Occupations in science-related fields can take place in many different settings
- Before you actually pursue a science career learn all that you can about yourself and the intended career by talking to future employers and gathering useful personal and career-related information
- It is important for you to make contact with the people that can help you grow in educationally appropriate ways
- Selection of the appropriate science career is a complex decision
- Never underestimate the power of the most important resource of all: The Human Resource
- The purpose of *Science Careers: Personal Accounts from the Experts* is to provide a helpful tool to assist science students and prospective science professionals in learning about select careers in the sciences

Suggested Reading List

Asher, D. *From College to Career: Entry-Level Resumes for Any Major from Accounting to Zoology*. 2nd ed. San Francisco, CA: Wet Feet Press, 1999.

Beatty, R. *The Perfect Cover Letter*. 2nd ed. New York: John Wiley & Sons, 1996.

Gottesman, D., & Mauro, B. *The Interview Rehearsal Book: 7 Steps to Job-Winning Interviews Using Acting Skills You Never Knew You Had*. New York: Berkley Publishing Group, 1999.

References

Carney, C., & Wells, C. *Discover the Career within You*. 4th ed. Pacific Grove, CA: Brooks/Cole Publishing, 1995.

Harmon, L. W., Hansen, J. C., Borgen, F. H., & Hammer, A. L. *Strong Interest Inventory. Applications and Technical Guide*. Palo Alto, CA: Consulting Psychologists Press, 1994.

Holland, J. L., Fritzsche, B. A., & Powell, A. B. *Technical Manual for the Self-Directed Search*. Odessa, FL: Psychological Assessment Resources, 1994.

Lincoln, Y., & Guba, E. *Naturalistic Inquiry*. Newbury, CA: Sage, 1985.

Myers, I. B., & McCaulley, M. H. *A Guide to the Development and Use of the Myers-Briggs Type Indicator*. Palo Alto, CA: Consulting Psychologists Press, 1985.

CHAPTER

2

Science Careers:

Statistical Exploration

Dr. Lamont A. Flowers and Dr. James L. Moore III

In keeping with the theme of this book, the purpose of this chapter is to present detailed information about science careers to give readers an idea of the number of persons pursuing degrees in science and engineering and the average salary for various science careers. This information is important for any person interested in learning about a career in the sciences.

This chapter is divided into seven sections. The first section discusses all the data sources used to prepare the graphs and tables in this chapter. The second section displays a graph showing the number of college degrees earned in science and engineering during the years 1991 to 2000. The third, fourth, and fifth sections, show graphs that display data on the number of persons receiving bachelor's, master's, and doctoral degrees in science and engineering fields, respectively. The sixth section presents data that shows the number of persons employed in various science careers. In the seventh section, we highlight the median annual salary for particular science careers. These data were presented to provide the reader with a statistical picture concerning the number of persons who are obtaining degrees in the sciences and how the numbers of individuals earning degrees have changed in recent years. We have also included a table that shows some differences in employment trends for select science careers.

Data Sources

To develop the graphs and tables that appear in this chapter, data and information on science careers were obtained from a variety of sources. The common theme among all of these data sources is that various divisions within the U.S. Department of Labor produced them. Below is a list of each data source used in this chapter followed by a brief description of each source.

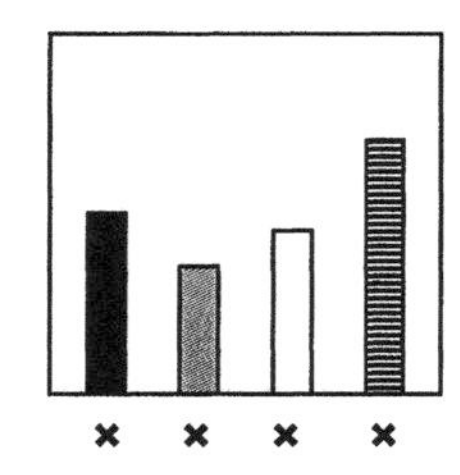

The goal of this section is to provide you with descriptive data on science careers and to inform you about those data that you should explore and become familiar with as you learn more about careers in the sciences.

Science and Engineering Degrees, by Race/Ethnicity of Recipients: 1991–2000
Contains a compendium of statistical information regarding the numbers of persons receiving college degrees in science and engineering by race or ethnicity and citizenship over a nine-year period.

National Industry-Occupation Employment Matrix
http://data.bls.gov/oep/nioem/empiohm.jsp
Displays the number of persons working in a given occupation and the expected growth rate or decline rate of the workforce for a given occupation. This website contains an easy-to-use database with information on more than 500 occupations.

Occupational Outlook Handbook, 2002–2003 Edition
http://www.bls.gov/oco/home.htm
Contains detailed career data including specific job duties and job qualifications, salary information, and future employment trends for a variety of occupations.

Degree Attainment in Science and Engineering

Data in figure 2.1 show the number of people who earned college degrees in science and engineering from 1991 to 2000. Of the degrees earned in science and engineering, most people earned a bachelor's degree during this time. Doctoral degrees were least likely earned during 1991 to 2000.

More specifically, in 1991, 356,785 bachelor's degrees were awarded in science and engineering. In 2000, 418,720 bachelor's degrees were awarded. In 1991, 72,828 master's degrees were awarded. In 2000, 88,191 master's degrees were awarded. In 1991, 24,023 doctoral degrees were awarded in science and engineering. In 2000, 25,979 doctoral degrees were awarded.

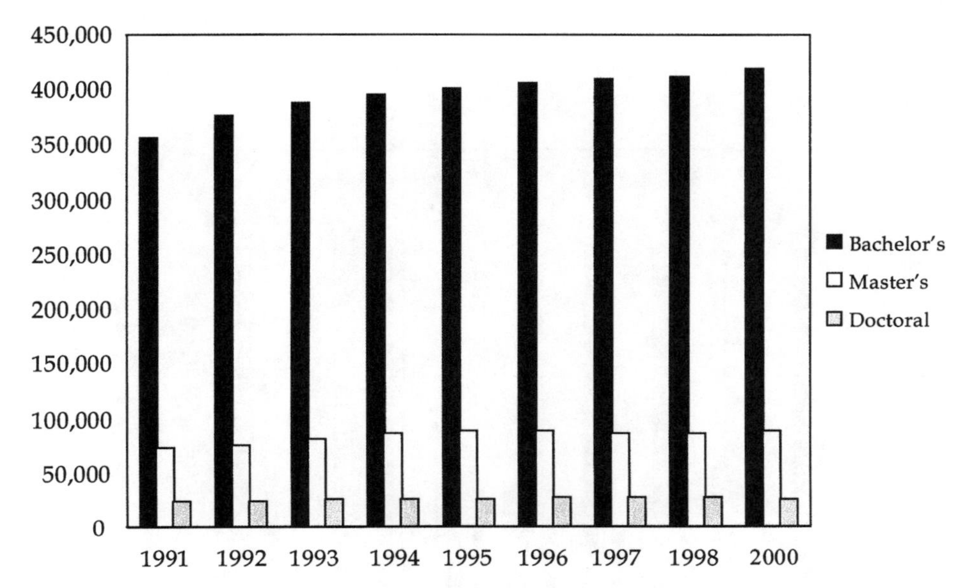

Figure 2.1 ***Number of Earned Degrees in Science and Engineering: 1991–2000***

Source: Science and Engineering Degrees, by Race/Ethnicity of Recipients: 1991–2000. National Science Foundation. No data were available for 1999.

Bachelor's Degrees in Science and Engineering

Figure 2.2 shows the number of earned bachelor's degrees in certain fields between 1991 to 2000. It should also be noted that you should consider obtaining and analyzing similar statistics for the science career that you are interested in pursuing. Thus, since these data in figure 2.2 represent a large number of careers, it is advisable to obtain additional statistics that are germane to a scientific or engineering career of interest to you.

As shown in figure 2.2, more bachelor's degrees were awarded in engineering fields than any other field. Bachelor's degrees in agricultural science were least likely awarded from 1991 to 2000. In 1991, 62,186 bachelor's degrees were awarded in engineering. Approximately 59,445 bachelor's degrees were awarded in engineering fields in 2000.

Biological science degrees followed engineering degrees in terms of the number of people who received them during 1991 to 2000. In 1991, 40,351 bachelor's degrees were awarded. In 2000, 64,800 bachelor's degrees were awarded in biological science. These data also showed that in 1991 approximately 16,407 bachelor's degrees were awarded in physical science fields. In 2000, 18,627 bachelor's degrees were awarded in physical science fields. Lastly, in 1991, 8,643 bachelor's degrees were awarded in agricultural science. In 2000, 18,486 bachelor's degrees were awarded in agricultural science.

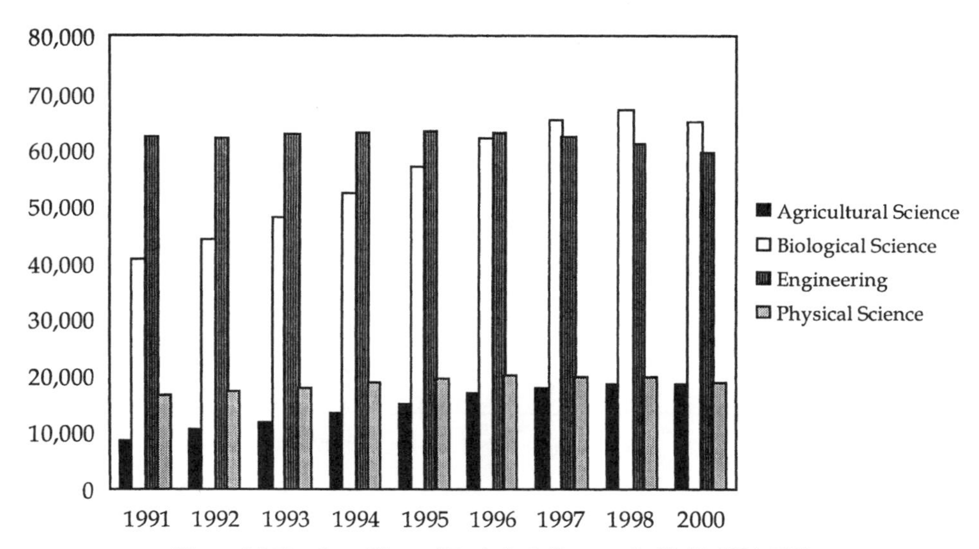

Figure 2.2 ***Number of Earned Bachelor's Degrees, by Field: 1991–2000***

Source: Science and Engineering Degrees, by Race/Ethnicity of Recipients: 1991-2000. National Science Foundation. No data were available for 1999.

Master's Degrees in Science and Engineering

Figure 2.3 shows the number of master's degrees earned in selected fields in science and engineering. As shown in figure 2.3, during the years 1991 to 2000, more master's degrees in engineering were awarded than in any other field. In 1991, approximately 24,007 master's degrees were awarded in engineering. In 2000, 25,722 master's degrees were awarded in engineering. Biological science degrees also accounted for a substantial percentage of the master's degrees earned during 1991 to 2000. In 1991, approximately, 4,806 master's degrees were awarded in biological science. In 2000, 6,228 master's degrees were awarded.

In 1991, 5,282 master's degrees were awarded in physical science. In 2000, 4,858 master's degrees were awarded. In 1991, 2,625 master's degrees were awarded in agricultural science. In 2000, 3,874 master's degrees were awarded. Overall, data in figure 2.3 show that most of the master's degrees earned during 1991 to 2000 were earned in engineering and biological science.

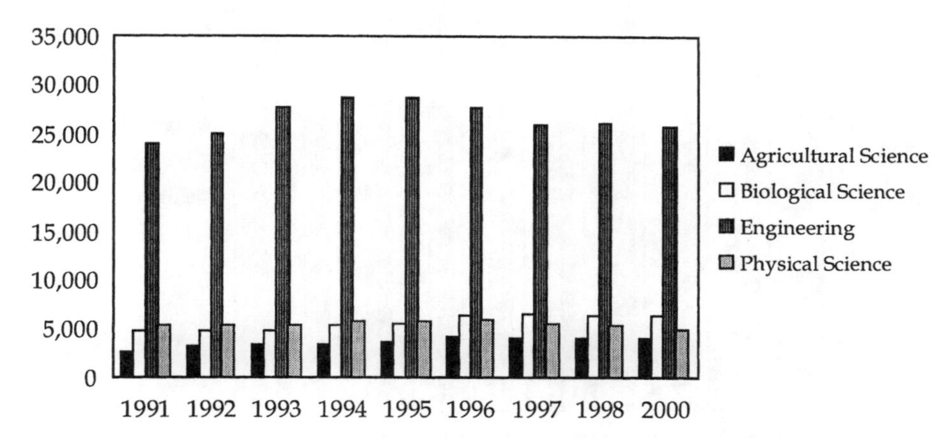

Figure 2.3 *Number of Earned Master's Degrees, by Field: 1991–2000*

Source: Science and Engineering Degrees, by Race/Ethnicity of Recipients: 1991–2000. National Science Foundation. No data were available for 1999.

Doctoral Degrees in Science and Engineering

Figure 2.4 displays the number of doctoral degrees earned in selected fields. As shown in figure 2.4, doctoral degrees in engineering were awarded more than any other field during 1991 to 2000. In 1991, 5,214 doctoral degrees were awarded in engineering fields. In 2000, 5,330 doctoral degrees were awarded. In 1991, 4,650 doctoral degrees were awarded in biological science. In 2000, 5,855 doctoral degrees were awarded.

Data in figure 2.4 also showed that, in 1991, 4,441 doctoral degrees were awarded in physical science fields. In 2000, 4,168 doctoral degrees were awarded. Doctoral degrees in agricultural science were least likely to be awarded during 1991 to 2000. More specifically, in 1991, 1,073 doctoral degrees were awarded in agricultural science fields. In 2000, 943 doctoral degrees were awarded.

Employment Data in Science and Engineering

The next section reviews employment data from selected science and engineering fields to provide you with information regarding the number of individuals currently working in these fields.

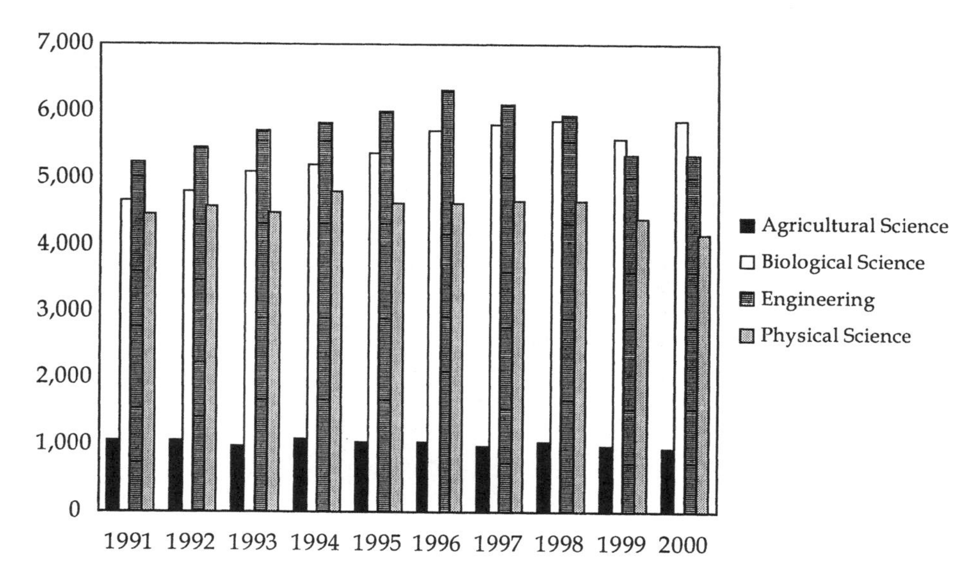

Figure 2.4 ***Number of Earned Doctoral Degrees, by Field: 1991–2000***

Source: Science and Engineering Degrees, by Race/Ethnicity of Recipients: 1991–2000. National Science Foundation.

Table 2.1 is based on employment statistics from the year 2000. While only a small number of careers are listed, the reader is encouraged to pursue additional information about other careers in science and engineering by examining the National Employment Matrix developed and maintained by the U.S. Department of Labor. As shown in table 2.1, registered nurses, who are employed in a large number of industries, had the largest number of workers among the occupations listed.

It is important to keep in mind that, while comparing these data are useful and can provide you with some helpful information, it may be more beneficial to make comparisons among the careers that interest you to get a better idea of how many persons are currently working in those fields. For example, if you are interested in pursuing a career in engineering but you are not sure if you would like to become a chemical engineer or an industrial engineer, then perhaps a more meaningful comparison for you would be to isolate these data for those specific occupations from tables 2.1, 2.2, and 2.3 to compare and contrast each specific field. Data from table 2.2 are also informative in that they show the expected change in percent of employment for select occupations by 2010. This information might be useful as you consider a particular science career to pursue.

EMPLOYMENT STATISTICS: 2000

Occupations	Number of Workers
Aerospace Engineers	50,434
Agricultural & Food Scientists	17,461
Agricultural Engineers	2,400
Astronomers & Physicists	10,063
Biological Scientists	73,101
Biomedical Engineers	7,221
Chemical Engineers	32,883
Chemists	84,323
Civil Engineers	232,046
Dentists	152,450
Dietitians & Nutritionists	48,740
Electrical Engineers	157,188
Environmental Engineers	52,421
Geoscientists	25,497
Industrial Engineers	153,636
Materials Engineers	33,491
Mechanical Engineers	221,443
Medical Scientists	37,006
Occupational Therapists	78,306
Pharmacists	216,865
Physical Therapists	131,822
Physicians & Surgeons	597,852
Registered Nurses	2,194,224
Veterinarians	58,634

Table 2.1 Employment Statistics: 2000

Source: National Employment Matrix, U.S. Department of Labor, Bureau of Labor Statistics.

As shown in table 2.2, the number of biomedical engineers is expected to increase by thirty-one percent over the next few years. Also, medical scientists are expected to increase by twenty-seven percent. Moreover, the number of environmental engineers is expected to increase by twenty-six percent.

Since the number of expected workers in any field depends on a number of factors such as the expected demand for their skills and services, expected

EMPLOYMENT TRENDS: 2000-2010	
Occupations	**Percent Change**
Aerospace Engineers	13.9
Agricultural & Food Scientists	8.8
Agricultural Engineers	14.8
Astronomers & Physicists	10.5
Biological Scientists	21.0
Biomedical Engineers	31.4
Chemical Engineers	4.1
Chemists	19.2
Civil Engineers	10.2
Dentists	5.7
Dietitians & Nutritionists	15.2
Electrical Engineers	11.3
Environmental Engineers	26.0
Geoscientists	18.1
Industrial Engineers	4.5
Materials Engineers	5.3
Mechanical Engineers	13.1
Medical Scientists	26.5
Occupational Therapists	33.9
Pharmacists	24.3
Physical Therapists	33.3
Physicians & Surgeons	17.9
Registered Nurses	25.6
Veterinarians	31.8

Table 2.2 Employment Trends: 2000–2010

Source: National Employment Matrix, U.S. Department of Labor, Bureau of Labor Statistics.

population growth over a given period, and advances in information technology these employment projection data shed light on which fields in science and engineering are expected to have the most jobs in the upcoming years.

Salary Data in Science and Engineering

To be sure, the salary that one might receive from one's hard work and persistence is a very important factor in the career decision-making process.

SALARY STATISTICS: 2000 (Median Annual Earnings)

Occupations	Salary
Aerospace Engineers	$67,930
Agricultural & Food Scientists	$52,160
Agricultural Engineers	$55,850
Astronomers & Physicists	$82,535
Biological Scientists	$49,239
Biomedical Engineers	$57,480
Chemical Engineers	$65,960
Chemists	$50,080
Civil Engineers	$55,740
Dentists	$129,030
Dietitians & Nutritionists	$38,450
Electrical Engineers	$64,910
Environmental Engineers	$57,780
Geoscientists	$56,230
Industrial Engineers	$58,580
Materials Engineers	$59,100
Mechanical Engineers	$58,710
Medical Scientists	$57,196
Occupational Therapists	$49,450
Pharmacists	$70,950
Physical Therapists	$54,810
Physicians & Surgeons	$160,000
Registered Nurses	$44,840
Veterinarians	$60,910

Table 2.3 Salary Statistics: 2000 (Median Annual Earnings)

Source: National Employment Matrix, U.S. Department of Labor, Bureau of Labor Statistics.

Stated differently, when deciding which career in science and engineering is right for you, it is wise to review salary information.

Table 2.3 reports the median annual earnings for selected fields in science and engineering. While the salary information reported in table 2.3 represents median annual earnings (which means that your actual salary could be either higher or lower than the salary reported in table 2.3), the information presented in table 2.3 is an excellent starting point to become familiar with analyzing salary data. As shown in table 2.3, of the occupations reported, aerospace engineers, astronomers and physicists, chemical engineers, dentists, pharmacists, and physicians and surgeons had the highest salaries.

Concluding Remarks

With the proliferation of computers, today's workforce is rapidly becoming more sophisticated and technologically advanced. The need for specialized skills and training to meet the present technological demands as well as future workforce projections is immense. In order to bridge your educational and occupational interests with employment trends in the 21st century, it is critical that you stay abreast with labor market information and resources. Such information has the propensity to provide insight and guidance for planning a career in science and engineering. In addition, acquisition of this information allows you to develop an occupational focus aligned with your interests, abilities, and aptitude.

In choosing an occupational path in science and engineering, it is usually recommended, if not required, that individuals have knowledge, skills, and aptitudes indicative for success in today's technological workforce. These prerequisites are fueled by the requests and sentiments of business and industry leaders. Their ultimate aim and pursuit is to find knowledgeable and skilled workers that can render profitable goods and services, at the lowest cost, for their companies.

Based on the workforce needs of business leaders and the companies that they represent, there are specific areas of competencies and training desired of prospective employees. Therefore, it is extremely important that individuals who aspire to enter as well as compete in this "new" labor market are mindful of these desired skills. The extent to which these individuals are successful in this burgeoning technological workforce is a function of their attitudes, beliefs, and educational experiences.

Take Home Points

- Research is the key to making good decisions about your future
- Review the *Occupational Outlook Handbook* and *Dictionary of Titles* (see the following website: *http://online.onetcenter.org*) to search for descriptions and titles of science and engineering-related jobs that interest you
- Seek out additional statistical information about a science career that interests you
- Many new jobs are expected in science and engineering in the upcoming years
- Keep a notebook or folder containing all of the information that you have collected about science careers
- Use print-material such as the *Wall Street Journal, Business Week,* and *Forbes* to keep abreast of your prospective career

Suggested Reading List

Peterson's Job Opportunities for Health and Science Majors: 1998. Edited by K. Hansen. Princeton: Peterson's Guides, 1997.

VGM's Careers Encyclopedia: A Concise, Up-to-Date Reference for Students, Parents, and Guidance Counselors. 5th ed. Editor. New York: Contemporary Books, 2001.

U.S. Department of Labor. *Report on the American Workforce.* 4th ed. Indianapolis, IN: JIST Publishing, 2000.

References

Hill, S. *Science and Engineering Degrees, by Race/Ethnicity of Recipients: 1991–00.* Arlington, VA: National Science Foundation, Division of Science Resources Studies, 2001.

U.S. Department of Labor. National Employment Matrix. U.S. Department of Labor, Bureau of Labor Statistics. Retrieved March 14, 2002 from *http://data.bls.gov*.

U.S. Department of Labor. *Occupational Outlook Handbook,* 2002–03 ed. Indianapolis, IN: JIST Publishing, 2002.

CHAPTER 3

Personal Accounts from the Experts

The major thrust of this book is to present helpful information about select science careers from science professionals from different scientific disciplines. The pages that follow are designed to provide the reader with a thorough, yet general description of the actual job duties of a particular science career. The goal of each contributor is to present a realistic view of his or her career and what it entails as well as to offer invaluable insights for a successful career to future students interested in pursuing a science career.

The science professionals that appear in this book offer sound advice, pertinent information, and useful recommendations and suggestions designed to benefit you, even if you are not quite sure you want to pursue the intended career. I urge you to let the life experiences and personal accounts from the experts highlighted in this book guide you as you consider a career in the sciences. The following science disciplines are addressed in this chapter: biotechnology, engineering, genetics, geology, medical technology, microbiology, molecular biology, nursing, pharmacy, physics, and virology.

BIOCHEMICAL ENGINEERING

> I never did anything worth doing by accident, nor did any of my inventions come by accident; they came by work.
>
> —*Thomas Edison*

Dr. Tonya L. Peeples

Associate Professor
Department of Chemical and Biochemical Engineering
University of Iowa

I. Biographical Sketch

With the example and encouragement of my parents, I learned how to find answers to my own questions at a very young age. I can remember as a child asking many questions of my parents and receiving the reply "I don't know, let's look it up." These experiences fostered my propensity for research, my interest in learning, and my desire to share information with other people. As a developing student, I excelled in science and math as well as in English. During the summer of my senior year in high school, I participated in a Student Introduction to Engineering Program at North Carolina State University (NCSU). I enjoyed the experiments in which I participated and decided to major in chemical engineering based on that experience. As a college student at NCSU, I obtained summer internship positions at Dow Corning in Midland, Michigan. For three summers I saw what professional chemical engineers were doing in technical service and support and in process development. While I enjoyed the engineering work, the best parts of my projects involved answering questions through designed experimentation. During my undergraduate study I developed interest in biochemical engineering. I studied biochemistry, biology, genetics, and biochemical engineering as part of my formal training. I was highly enthralled with the study of biochemical engineering. Biochemical engineering involves the study of the various chemical processes that occur in biological systems. In the summer after my senior year, I participated in a Research Experiences for Undergraduates (REU) program working on a monoclonal antibody project. Although, I was offered a job at Dow Corning, I opted to pursue further opportunities to do biochemical engineering research by going to graduate school. After obtaining my bachelor of science in chemical engineering with a bioscience option, I entered Johns Hopkins University (JHU) as a chemical engineering graduate student. My graduate work, under the guidance of professor Robert Kelly, involved the study of organisms known as archaea, which represent a new domain of life. After obtaining my Ph.D., I worked as a postdoctoral researcher in environmental engineering science at the California Institute of Technology with professor Mary Lidstrom. I studied the genetics of a bacterial membrane protein, methane monooxygenase, which is important in cleaning up environmental pollutants. In 1995 I joined the faculty of chemical and biochemical engineering at the University of Iowa as an assistant professor. The breadth of experiences as a graduate student and postdoctoral fellow has afforded me the opportunity to develop a unique research program in extremophile, interfacial, and environmental biocatalysis. Biocatalysis is the use of living systems or the components of living systems to carry out chemical reactions.

My research team studies environmentally and medically relevant bacteria and fungi as well as "extremophilic" microbes. Extremophiles may live at temperatures near boiling or under the high pressures of the deep sea, in the presence of high salt or in highly acidic or alkaline environments. We are evaluating the unique biochemical features that have enabled these organisms to adapt under harsh chemical conditions. In general biochemistry is the study of the chemical mechanisms of living organisms. Understanding the biochemical functions of extremophiles may help us to increase the stability of biological systems that are used in everyday applications. We use knowledge of molecular biology, classical cellular physiology, and bioprocess design as tools of discovery. These research efforts have been well funded and have provided great learning opportunities for graduate and undergraduate chemical engineers, biologists, microbiologists, and biochemists. I received several awards based on my research and teaching including a General Electric Faculty for the Future award, a National Science Foundation CAREER award, a UI Collegiate Teaching Award, and an invitation to participate in the National Academy of Engineering's "Frontiers in Engineering" conference. I have also enjoyed significant funding from private companies as well as federal agencies. I have published fourteen publications reporting cloning, isolation, and characterization of extremophilic enzymes, characterization of extremophile components, fungal biocatalysis, and aqueous two-phase reactor systems for biocatalysis. I was promoted to the rank of associate professor with tenure in 2002. I am a member of the American Chemical Society, The American Institute of Chemical Engineers, The Society of Women Engineers, The American Society of Engineering Educators, and The American Association for the Advancement of Science.

II. Description of Job Duties

The job of a professor is to participate in the activity of scholarship through developing and maintaining an active research program, teaching graduate and undergraduate students, and serving society. Professors with active research programs are constantly increasing their knowledge and advancing the field. They are able to teach the latest information to both graduate and undergraduate students. Through service activities, professors also enable institutions of higher learning and are also charged with improving the quality of life for the citizens of their home states, the nation, and the global community.

Research

In advancing my career as a biochemical engineering professor, I have set a primary goal of developing a unique, nationally and internationally recognized research program in biochemical engineering. The specialty areas of extremophile biotechnology, interfacial biocatalysis, and environmental microbiology captured my imagination during my graduate career and have sustained my intellectual curiosity. Along these lines, research projects in my laboratory have laid a foundation for research of lasting impact on the biochemical engineering discipline. To pursue these areas of research, I have written many proposals. In a typical proposal, I give general background information, identify a problem that must be solved, and state a strategy for solving the problem. A successful proposal must convince a funding agency or company that the project is important. It must also demonstrate that I have the qualifications and the creativity that will enable me to successfully complete the activity. Once research funding is obtained, I must recruit talented people to help me carry out the project. In these efforts, I have had several graduate students and postdoctoral fellows join my research team. In this capacity of research mentor I must help these students and fellows develop research skills required to complete their portions of the project. I also have to purchase the equipment and supplies to do the research and make sure that the people in my lab have everything they need to get the work done. As my research team makes discoveries and completes projects, we have to publish the information to inform the community at large. Publishing our work comes in the form of writing technical papers that are reviewed by other scientists, making presentations at scientific meetings, and writing patents to seek commercial benefit from truly new technologies.

Teaching

Instilling enthusiasm for using science and math is the task of a teacher in a vital engineering program. This program is made dynamic by encouraging students to take what is learned in the classroom and apply it to real engineering problems. As a research mentor, an academic advisor, and a teacher, I have been able to inspire students to develop a deeper understanding of engineering and a broader view of the engineering problems to be faced in the future.

The general course load for a full-time faculty member in the UI College of Engineering (COE) is three courses per year. One course generally consists of three hours of in-class time per week. In a typical course, a professor spends two hours in preparation for every one hour of in-class activity. Beyond preparing in-class materials the instructor is responsible for developing the course learning goals and devising a strategy for enhancing and assessing student learning. I select and review materials such as textbooks, develop lesson plans, lecture notes, homework problems, exam problems, laboratory experiments, and design projects. Within the UI COE, professors spend most of the contact hours with students. Teaching assistants (typically graduate students) help us grade homework assignments and answer questions outside of class.

Today's students who graduate from chemical engineering are directed to diverse fields including pharmaceuticals, materials science, environmental management, technology transfer, law, medicine, and business as well as traditional chemical and petrochemical industries. I encourage students to develop a depth of understanding of engineering problem solving complemented by a breadth of exposure to engineering problems. I have provided a diversity of learning experiences by expanding laboratory opportunities for students, adding open-ended design projects to courses, including discussions of topical issues in coursework, and fostering individual investigations through research opportunities in my laboratories.

Since I have an active research program in the field of extremophiles, developing a course in the "Biotechnology of Extremophiles" seemed a natural extension of my efforts into education. This course addressed the application of biological systems from extreme environments. Students reviewed the evolution and engineering of biocatalysis under extreme conditions. This type of activity is required of faculty because it enables professors to truly inspire students.

Service

As a faculty member at the University of Iowa, I am called to give of my time and talents to my local community, my department, my college and university, and to the discipline of chemical engineering. One of the focal points for many of my service activities has been the promotion of diversity. As a teacher and role model in engineering, I am challenged to improve the graduate and undergraduate environment for female and minority students in chemical engineering and engineering at large. Through professional development programs on campus and in engineering organizations, I am helping to foster an environment that encourages engineers of different backgrounds to recognize their common interests and goals. This issue is met by improving the appreciation for diversity in both majority and minority groups. If professionals and scholars recognize common goals of excellence in engineering work, differences and the aversion to differences can be erased as obstacles to success.

Other service activities have involved providing technical assistance and advice from a biochemical engineer's prospective to private, professional, and governmental constituencies at the local, state, and national levels. My additional service activities include tutoring 6th through 12th grade students in math, reviewing technical papers and proposals on federal and state panels, and serving on committees at the department and university level.

III. Advantages of Career

Biochemical engineering is an exciting discipline that increases our understanding of the physical world. Skills of engineering problem solving are applicable to a wide variety of fields including petrochemicals, foods, the environments, business, microelectronics, polymers, pharmaceuticals, and medicine. Because of the adaptability of biochemical engineering training to a variety of fields, biochemical engineers draw high salaries at all levels of education. It is common for undergraduates to receive "real" salaries for summer internships that can be used to pay some of the college expenses. There is also a reasonably high salary for biochemical engineers who choose to work in industry after obtaining the bachelor of science degree. Most of these engineers work in manufacturing, process engineering, and technical sales and support areas. Graduate education offers unique opportunities to participate in new discoveries. Biochemical engineers who are interested in research and development are the ones who pursue graduate degrees. Most graduate programs pay engineering students as teaching and research assistants while they are completing their studies. Engineers with master's and doctoral degrees must then choose between private industries, national laboratories, or university settings for their career. Working in the academic setting allows flexibility in the number and types of problems that can be studied. There is the potential to make an impact in new areas of discovery. There is also the ability to make an impact on developing young engineers. In terms of lifestyle, the work hours are flexible and the opportunities for advancement are clearly connected to scholarly productivity.

In biotechnology, unique chemical engineering skills for process design and analysis have enabled engineers to make an impact in the characterization and application of biological systems. Even with the greatest skills in biochemical engineering, most contemporary problems still require cooperation within the scientific and engineering community. Being at a university enables this type of collaboration. Through interdisciplinary efforts with colleagues in microbiology, biochemistry, medicinal chemistry, chemistry, food science, and geological science as well as other engineering disciplines, I have been able to develop innovative research and graduate training programs. This has also increased the opportunities for me to learn new things.

IV. Essential Related Skills for Success

The tools that will enable engineers to be effective in the global community include: 1) the ability to synthesize and deconstruct complex problems, 2) the ability to research and find answers to original questions, and 3) an ability to communicate new discoveries effectively. Technical competency in core engineering areas is a must. However, there are also interpersonal and professional skills that go beyond the science and engineering. Being able to find the right answer is important, but it is not the end of solving the problem. An engineer must be able to market ideas and make an impact on the community.

Undoubtedly, gaining technical skills will enable students to become good problem solvers. In the process of solving problems, a practicing biochemical engineer must have a solid understanding of basic math and science, a foundation of biochemical engineering core areas of thermodynamics, transport phenomena and reactor design, and a basic understanding of process economics. Beyond engineering skills, a practicing engineer must also be able to work well in teams and to explain technical concepts through writing and public speaking. Any accredited biochemical engineering program will seek these outcomes for its graduates.

Beyond technical skill, successful engineers must have initiative and motivation, strong engagement skills, leadership skills, an awareness of diversity, and the ability to continue learning through the span of their careers. Many of these skills can be fostered even before a student enters college and graduate school. In my case, involvement in extracurricular activities (such as the school newspaper staff, the math club, the debate team) enabled me to enhance my abilities for leadership, teamwork, and communications. Developing a desire to learn was also something that was valued in my household. Respect for others different from myself was also a value that was taught at home, in my local church, and in the schools I attended. These values were also passed on to me from various mentors who touched my life during elementary and high school, in college, in graduate school and within my professional environment. These are values that I am now able to share with students and professional colleagues and my local community. As an undergraduate, I continued extracurricular activities and completed professional internships. These activities enabled me to develop skills at time management, experimentation, and organization. As a graduate student, I was able to learn more about managing a research project, communicating technical information in both written and spoken form, and mentoring students. These are skills I use today as a chemical engineering professor to balance teaching, research, and service activities as well as my personal life.

V. Additional Advice

It is important as students pursue their lifelong careers to appreciate and acknowledge those people who have acted as mentors on the journey. I honor the contribution of many mentors in my career by extending advice and encouragement to the new students who are just beginning their own journeys. Even as an aspiring biochemical engineer or professor, you are never too young to be an encouragement to someone else. Service is the highest form of contribution to improving the world. Actively seek out mentors who can give you feedback about your activities and advice about developing your professional goals. Mentors can be found within student organizations and professional societies, within a local religious community, within your family, and among your faculty. Sometimes mentors will find you. More often you must express some interest in seeking advice.

This is a place where you can demonstrate initiative and motivation. Do not be afraid to ask questions in class. Take advantage of a professor's office hours. Inquire about research and employment opportunities on and off campus. As you pursue internships you will be able to survey what people in your chosen field are doing in their jobs on a daily basis. This experience is valuable as you formulate your career goals. It also gives you opportunities to find industrial mentors and to improve your resume. Don't be afraid of being a "nerd." Find answers to questions that arouse your curiosity. Read about new discoveries and technological developments. These are activities in which research professors and engineers and scientists participate on a daily basis. Talk about science. Share your enthusiasm for science with others. In facing the complex problems of the future, more people need to be scientifically literate. Science literacy enables society to make informed decisions about proposed solutions. Learn how to work with others. Seek out study groups. Participate in clubs, professional organizations, and athletic teams. These activities will enhance your leadership skills, especially your abilities to engage, cooperate with, and influence others. Biochemical engineers have unique skills of process design and evaluation, which have enabled us to make an impact in characterizing the nature of biochemical systems. Still most contemporary problems require cooperation within the scientific and engineering community. For aspiring engineers it is important to develop a breadth of experiences to facilitate working effectively in multidisciplinary teams. Here the same tools that are used to handle diversity of gender, ethnicity, and race in clubs and teams can be applied to handle differences in academic background and professional experience in professional settings.

VI. Pitfalls to Avoid

Avoid being distracted from your goals. Choose your activities wisely so that you can give your best performance in your coursework and on your job. Difficulty balancing school demands as well as extracurricular activities caused me to struggle in some key courses. I could have performed better, but I persevered and learned a lesson about how many activities I could handle at one time. This lesson helps me as a faculty member when I am considering service activities. I respond to requests that are compelling and avoid overloading with too many extra jobs. When you do struggle, do not lose heart. Be persistent at doing your best to reach your goals. Persistence pays off when learning difficult concepts, during grant writing, during challenging experimental investigations, and when mentoring students. Seek feedback and support from research program directors, mentors, and trusted colleagues.

Keep a positive attitude. A key point to dealing with difficult professional situations and relationships is to remain constructive. As you continue to give your best, you honor those mentors who are supporting you. Avoid negative self-talk. You can overcome external negativity with positive thoughts and actions. The most challenging time in my career came during my third year in graduate school when both my father and brother died. My career was immediately put into perspective. Nothing is more important than family. The positive impact that my brother and father made in my life carried me through. I learned that my engineering career could be gratifying but that my true goal is to share my accomplishments with others significant to my life. Avoid shyness. Don't be afraid to boast about your achievements. Promote yourself and establish a visible professional presence. Publish your research findings quickly. Encourage your students to think about the presentation of their work even as they design their experiments. Get some training in management to help you in handling research budgets, laboratory equipment, personnel, and time. This is something that is not covered extensively in graduate education, but is an important skill in becoming a successful professor. A final word of advice is to avoid burning out. Remember to pay attention to your emotional, physical, and spiritual well-being. An academic biochemical engineering career can demand a large amount of time and energy. Anxiety and exhaustion can take over if you do not make time for family, good nutrition, exercise, and spiritual renewal.

VII. Qualifications Required

To obtain a tenure track faculty position in a biochemical engineering department, a Ph.D. in chemical engineering or similar field (physics, chemistry, biochemistry) is required. Some researchers also do postdoctoral work to gain experience for their research careers, but this is not strictly required. In order to get a Ph.D. a student must complete a bachelor's degree in chemical engineering or a related field and then enter graduate school.

For achievement of tenure, a faculty member must establish a professional reputation for scholarship, through getting research grants funded and publishing articles in peer-reviewed journals. The faculty member must also be a good teacher and participate in service activities.

(i)nteresting Facts

❶ College and university professors work with approximately 15 million students each year.

❷ Additional information on the qualifications and preparation required for a career as a science professor can be obtained from: *http://www.aacu-edu.org*.

❸ University faculty typically have flexible schedules.

Source: U.S. Department of Labor. *Occupational Outlook Handbook*, 2002–03 ed. Indianapolis, IN: JIST Publishing, 2002.

Take Home Points

- A biochemical engineering professor participates in the activity of scholarship through maintaining an active research program, teaching students, and serving society
- Biochemical engineers draw high salaries at all levels of education because of their adaptability; the ability to solve biological problems is an example of this versatility
- Successful biochemical engineers are problem solvers with initiative, strong engagement and leadership skills
- Aspiring biochemical engineers should actively seek out mentors, develop their independent learning skills, and learn teamwork through participation in extracurricular activities
- In pursuit of academic careers, avoid distraction, burn out, shyness, and negativity through being positive, persistent, and keeping things in perspective
- B.S. and Ph.D. degrees in biochemical engineering or a related discipline are necessary to become a biochemical engineering professor

Suggested Reading List

Feibelman, P. *A Ph.D. Is Not Enough: A Survival Guide to Science*. Reading, MA: Perseus Books, 1993.

Roberts, R. *Serendipity: Accidental Discoveries in Science*. New York: John Wiley and Sons, Inc., 1989.

Wankat, P. *The Effective, Efficient Professor: Teaching Scholarship and Service*. Boston, MA: Allyn and Bacon, 2002.

BIOTECHNOLOGY

> If I can't dance, I don't want any part in your revolution.
>
> —*Emma Goldman*

Dr. Amy L. Springer

Biotechnologist
Senior Scientist
Prolinx, Inc.

I. Biographical Sketch

As a child I was always interested in nature and animals. My parents are both scientists and they encouraged me to be curious about the world around me. My early education was largely in alternative schools, whose classroom structures were flexible and where I took advantage of opportunities to go on nature walks, visit museums, or read about animals. I knew I wanted to study genetics the moment I read about the genetic basis for animal fur color. I was fascinated by the idea that it was possible to determine what genes were present by analyzing the fur colors of parents and their offspring.

In high school, I enjoyed all of my science classes, but biology was the class I enjoyed the most. When it was time to decide which college I would enter I analyzed several factors. Principally, I wanted to attend a small school so I could have opportunities for more contact with the teachers. Moreover, I wanted to attend a liberal arts school because I was interested in many subjects, but I also wanted the school to be strong in the sciences with opportunities for undergraduate research. After careful analysis I finally decided to attend Mount Holyoke College. Mount Holyoke College is a women's college with a rich history in promoting scientific study for women.

Throughout my undergraduate education, I was immensely fascinated with the molecular mechanisms of genetics and biochemistry so I decided to purse a graduate degree in molecular biology at Princeton University.

In graduate school I learned much more than the specific subjects I studied, I had to learn how to apply myself, to understand subjects in detail, and how to effectively communicate to others what I understood. I loved the process of deducing what happened in the wild type (normal) organism based on what happened when you mutate a gene. After graduate school, I moved to the California Institute of Technology where I engaged in post-doctoral work in environmental molecular microbiology. In this field the goal is to apply information learned from studying model organisms and their genes in order to address a particular environmental issue, such as improving waste water treatment processes. When it came time to find a permanent position, I wasn't sure where to go. In my studies I was enamored with the elegant early genetics experiments with viruses that infect bacterial organisms (bacteriophages) and fruit flies, but I felt like the field of biology was changing from one where elegance and inference ruled to one where new technologies enabled scientists to obtain information by "brute force." After a thorough job search I decided to join the world of cutting-edge biotechnology in private industry. Biotechnology generally involves the application of knowledge of certain biological processes of living systems or existing technologies in order to accomplish various industrial or commercial goals that have a clear economic and social benefit. For example, one of the first genetically engineered systems involved the production of human insulin for medical applications.

I am currently a senior scientist in applications development at Prolinx, Inc., a biotechnology company in Bothell, WA. I am part of a research and development team that develops the company's proprietary technologies into molecular biology research tools. I am also involved in many volunteer activities in my community and serve on the board of the local chapter of the Association of Women in Science.

II. Description of Job Duties

Research

At Prolinx, Inc., I have been involved in many projects to develop innovative tools for molecular biology and genomics research. The primary technology on which this company is based is a small-molecule chemical affinity system for specifically capturing or extracting proteins and nucleic acids. This technology makes it possible for scientific researchers to isolate specific proteins or nucleic acid molecules present at very small amounts for further study. This technology is useful for various high-throughput molecular biology applications. In addition, I was very instrumental in developing our first product, a novel magnetic–particle–based dye removal system for DNA sequencing analysis and other analyses using fluorescently labeled nucleic acids. DNA sequencing is the process of determining the exact composition and order of the chemical constituents (nucleotides) that make up a DNA molecule. Currently, I am working on a surface plasmon resonance–based biosensor system for studying interactions of biomolecules. This biosensor will enable the study of interactions, such as those between proteins or between proteins and small molecule effectors, at a very sensitive level. Applications research is, as the name implies, very applied. My projects involve taking a technology platform whose properties we know, and finding ways to adapt it to perform particular functions that are useful to laboratory researchers. Examples of these are purification of unincorporated dye terminators from a DNA sequencing reaction or purification of a protein from an affinity column specific to that protein. An application will be successful only if it can perform this function more easily or less expensively than technologies that already exist. An important area in biotechnology is the development of tools for performing standard molecular biology or biochemistry assays in very small volumes and at high throughput. Applications development research is very focused on the research goals of the scientific community. Stated simply, we develop processes that make it easier for scientists to perform their research.

Working at a small innovative company has given me the opportunity to keep abreast of forces driving biotechnology and have broad exposure to new technologies. I work jointly with scientists and engineers from other disciplines and need to be able to adapt quickly to new projects and ideas. Developing tools for biotechnology that apply state-of-the-art technologies generally require a joint effort, and I have had to establish collaborations in both academic and industrial settings, seek out appropriate resources for new technologies, and develop new project proposals.

Management Responsibilities

Management responsibilities include directing research teams consisting of two or three individuals, developing project plans, conducting interviews and making hiring decisions. Other responsibilities include evaluating and making decisions or recommendations about purchases of equipment or reagents.

Training and Outreach

I also assist in the training of new scientists or new marketing staff members. Much of my work involves describing our technologies to scientists outside the company. It is therefore imperative that I understand the research goals of other scientists to know how our technology might help them. I am often involved in discussing potential collaborative studies and writing technical reports or proposals for new products, and have been the principal author in peer-reviewed journal articles describing our technologies. Other responsibilities include being available to answer technical questions from customers, and representing the company at scientific meetings several times throughout the year.

III. Advantages of Career

Small biotechnology companies are exciting places to work, the pace is rapid, and there are always new technologies to be learned. Smaller companies usually offer their employees more opportunities to learn new skills and to advance quickly. Because of more limited resources, smaller companies often enter into collaborations with other companies, academic institutions, and customers. These activities provide lots of opportunities for interacting with other people and for the development of good communication skills. Lastly, it is very rewarding to contribute to new scientific technologies that improve aspects of research processes. These technologies contribute to the overall quality of human life.

IV. Essential Related Skills for Success

A career in biotechnology utilizes all the basic scientific skills such as analytical reasoning, creativity, critical thinking, and keen observational skills. Having good communication and interpersonal skills enables one to work well as part of a team or in a management role. One should be able to work independently and be comfortable in an interdisciplinary environment. One important skill, I believe, is the ability to set priorities that balance immediate and long-term goals and maximize use of available resources. Some understanding of economics and business development are important for advancement in this type of career, and can often be acquired on the job through utilization of good observational skills.

V. Additional Advice

When considering joining a biotechnology company, evaluate not only the project you will work on initially, but also the company as a whole. Do you like the people with whom you would work? How do these people seem to get along with one another? Is your immediate supervisor someone you think you can communicate with well? What opportunities for advancement would exist for you? Also, find out about the company's history and its sources of funding. Do they have a sound business plan, one that seems likely to keep the company viable or even allow it to grow? A company with a bleak future can turn into an unpleasant place to work. Does the company management seem to value the contributions of its scientists? If not, this can lead to morale problems. Choosing a positive work environment is important for enabling the smooth advancement of your career. If, after you have been in a job for a few years, you believe your work is becoming too routine or even unpleasant, don't be afraid to try something new. You will do your best work if you enjoy what you are doing. It is not uncommon to stay in any one job in biotechnology for only a few years and then move on to do something else. This is a great way to learn new skills or develop old ones. I think that I may someday return to a more basic research environment, perhaps an academic one, to apply some of the genomics tools I have learned as a biotechnologist.

VI. Pitfalls to Avoid

Communication skills are important in any position; if you are having problems at work, seek ways to make sure that you are able to communicate your issues to the relevant people and, just as important, make sure you can listen effectively. Most problems are solved simply when two sides learn to understand each other's perspective.

If you are in a position to supervise others, make sure you foster an environment in which your staff feels comfortable communicating key issues. It is important to understand that different people have different working styles (particularly when they are different from your own), and that they will do their most productive work when they are most comfortable. Finding the balance between motivating an employee and allowing for his or her personal style is perhaps the most challenging aspect of a supervisor's role.

VII. Qualifications Required

Some scientific background is generally required to work at a biotechnology company, but a Ph.D. is not a requirement. Typically, a bachelor's degree in a science-related field is required for a career in biotechnology. A graduate degree will usually allow you to begin your career at a higher pay scale and level of responsibility within a particular biotechnology company. However, I have known many career biotechnologists who first started in the wash room or receiving room and worked their way up to responsible positions within the company. The skills you do need include the ability to learn new technologies or biological systems, to apply the scientific knowledge you do have, and to prioritize information.

(i)nteresting Facts

❶ Biotechnology industries improve technologies in human medicine, agriculture, basic research, industrial processes, and veterinary medicine.

❷ The biotechnology industry will experience a surge in employment in the next decade due to continuing scientific advancement.

❸ Biotechnologists can work in many different settings.

Source: U.S. Department of Labor. *Occupational Outlook Handbook,* 2002–03 ed. Indianapolis, IN: JIST Publishing, 2002.

Take Home Points

- Applications development biotechnology involves using innovative new technologies and developing tools that make it easier for researchers to perform their experiments
- Biotechnologists need to be able to explain their company's technology to other scientists and be able to learn about other scientist's research
- Enjoy the process of doing science, of solving the puzzle. A career in biotechnology involves finding solutions to a wide range of problems
- Don't be too specialized. Biotechnologists need to able to learn new skills and switch projects adeptly
- In order to work productively in your work environment, and to represent your company's technology to others, communication skills are very important
- A bachelor's degree in a science-related field is required for a career in biotechnology. A graduate degree will usually allow you to begin your career at a higher pay scale and level of responsibility within a particular biotechnology company

Suggested Reading List

Brown, S., & Rowh, M. *Opportunities in Biotechnology Careers.* Chicago, IL: VGM Career Books, 2001.

Fisher, R., Ury, W., & Patton, B. *Getting to Yes: Negotiating Agreement without Giving in,* 2nd ed. New York: Penguin Books, 1991.

Keller, E. *A Feeling for the Organism.* New York: W. H. Freeman and Co., 1983.

ENGINEERING

> An education isn't how much you have committed to memory, or even how much you know. It's being able to differentiate between what you know and what you don't.
>
> —*Anatole France*

Mr. Toby Q. Jenkins

Industrial Engineer
Liquids Organizational Effectiveness Manager
Proctor & Gamble Manufacturing Company

I. Biographical Sketch

As a kid, I was often getting into trouble for taking appliances apart. I indeed had an unquenchable thirst to know how things worked. How do they make rubber? How does cement stay hard? What keeps airplanes in the air? Could we live on Mars? In school, I always enjoyed math and science. Both subjects fueled my curiosity of how things work. My junior year of high school, I started thinking seriously about the most appropriate major in college for me. Although I didn't exactly know what to choose I knew that my ultimate selection needed to involve a field that most aptly correlated with my interests, academic strengths, and goals. The best advice regarding major selection came from my brother, who is six years older and was majoring in architectural engineering. He introduced me to the world of industrial engineering. Following several meaningful discussions with my brother regarding the field of industrial engineering I did some additional research on my own and found industrial engineering to almost be perfect for me. The discipline of engineering can best be described as the application of science and mathematics by which various properties of raw materials are made useful for a particular purpose or outcome. Industrial engineering is a combination of the economic basics of business and the analytical and technical aspects of mechanical engineering into one curriculum. On the technical end of industrial engineering, you can specialize in ergonomics, automation, facility design, or material handling. On the less technical side, industrial engineers can specialize in engineering economics, organizational design, or quality control. Industrial engineers provide services or facilitate the production of materials by ultimately determining the most efficient means of utilizing raw materials, machines, and personnel. To accomplish these goals industrial engineers make use of sound management theories, mathematical models, and common sense. It is one of the most versatile disciplines of engineering. Industrial engineers are in demand in many fields such as industrial insurance, financial banking, manufacturing, construction, and the defense industry.

As an undergraduate I received several academic honors in engineering. Also during my undergraduate career I participated in highly educational internships at IBM and General Motors. These wonderful experiences gave me the unique opportunity to learn more about industrial engineering. Also early in my career I was fortunate to work with a quality mentor who not only taught me how to more precisely perform my job duties but also provided me with excellent advice, guidance, and counseling that has proven to be very instrumental during my career.

II. Description of Job Duties

I have worked for Procter and Gamble (P & G) for almost a decade in various manufacturing roles. At Procter and Gamble the scope of job duties for an industrial engineer is very broad. Within my tenure I have managed construction products, supervised the production of several products, and managed human resources. In my current role, I am responsible for the packaging of such products as Mr. Clean, Febreze, and Cascade. In this role I have a team consisting of fifty-three persons who assist me in meeting the production goals for our manufacturing plant. The challenge in this role is to stretch the capability of my team, manufacturing systems, and resources to package quality goods as safely and cheaply as possible. The ultimate measure of our success is determined by the production cost of a particular product. My goal is to decrease the cost of manufacturing a particular product. Every year, the expectation is for the production cost to decrease, usually in the range of five–ten percent per year. Production cost consists of labor, materials, equipment, and logistics cost. The primary means of managing the aspects of production cost is through developing and monitoring every process involved in producing various products. My job is to make sure that every aspect of production is efficient, accurate, and precise.

Maintaining this kind of competitive edge has become increasingly challenging and complex in recent years with the development of the global economy and the recent recession. These events have put a premium on profitability. Domestic P & G facilities, like many other U.S. companies that operate throughout the world, have to be more productive than their international competition to compete with cheaper wages in other countries. However, domestic facilities can't necessarily invest in the latest technology and automation to gain this productivity edge. Ultimately, the bottom line is generating shareholder returns. Shareholders refer to individuals who hold shares of stock in Procter and Gamble. The ultimate measure of success of the company is driven by the value or return the public gets from owning P & G stock. At my level, I impact P & G stock price by exceeding budget expectations and generating additional savings in running my business.

III. Advantages of Career

The most significant advantage of my career is the wide range of job duties that are encountered. When I was interviewing for jobs as a college senior, it was a definite turn off to interview with engineers who had done the same job for ten, twelve, eighteen years. The variety of job duties allows me to gain new and exciting competencies outside the field of engineering. My career also presents me with the opportunity to interact with people from many different educational backgrounds. Another advantage of my career is the constantly changing work environment. To continue to be the best consumer products company in the world, P & G is constantly pushing its systems, employees, and manufacturing processes to be better. Just like championship caliber professional sports teams who never stop seeking to improve, P & G operates in the same manner by constantly upgrading systems, and providing continuing educational opportunities for its employees to meet the demands of a changing technological business world. Another advantage of being an industrial engineer for a manufacturing company is the satisfaction you receive when you meet the production projections for a particular product.

IV. Essential Related Skills for Success

I cannot overemphasize the importance of being well rounded: being able to communicate effectively, relate well to other people, and perhaps the most important, possess quality leadership skills. If you choose to use science in a business setting the way I have, these other skills become very important. I picked up these skills in school through playing sports and serving in leadership roles with various organizations from student government to community service organizations and by taking several courses in college.

In terms of my own education I found my technical writing class to be particularly important. Before I took this very informative class I was not aware of the rules and procedures for writing effectively for professional settings, such as progress reports to management, interdepartmental memos, etc. This class helped me to develop the necessary skills that an industrial engineer needs to effectively present the data in a style and form accepted by the professional community. The goal here is to understand that in addition to a comprehensive knowledge of engineering and mathematics and related scientific disciplines you also must possess other important skills that will ensure a successful career.

Engineers that are "book-smart," are a dime a dozen. Those that rise to the top are able to take their book smarts, influence and lead others, and effect positive change. A good idea is a waste if it cannot be communicated effectively. In my job, I use my analytical background and common sense to make decisions and decipher data, but the leadership skills and interpersonal skills often are more important in persuading others to action.

V. Additional Advice

Know yourself! Specifically, know where your passions are and are not. Money can make you happy temporarily, but doing what makes you happy and what you are passionate about will bring you happiness forever.

Internships are the bridge between academic and professional institutions. Internships allow you to gain practical experience, provide professional development, offer the potential for future employment, and may provide financial compensation. If you are a high school or undergraduate student I recommend that you seek out an internship at a local industry or your current university. This will give you the opportunity to learn more about the work environment.

Internships, summer programs, and independent study courses at your university can provide you with the opportunity to learn a wide variety of useful skills that will prove invaluable as you begin your career. The ultimate goal for this type of introductory experience is to simply learn, in a general way, the basic job duties and educational competencies needed for a successful career in industrial engineering so that you can apply what you learn later down the road. Based on my own experiences I have learned that it takes time to develop these competencies and related career-success skills. I am suggesting that you begin the learning process as soon as possible so that when you assume a professional position your transition from student to professional will be a smooth one.

There are a number of ways to find out about potential internship opportunities including the Internet, counselors, advisors, teachers, and professors. Once you have selected the internship that you think will be best suited for you the next step is to make contact (e.g., telephone, e-mail, standard mail) with the institution granting the internship. Use this opportunity to find out whether the internship will benefit you.

VI. Pitfalls to Avoid

The major pitfall to avoid when pursuing a career in industrial engineering is becoming distracted from your goals. It is important that you stay focused during matriculation in your engineering undergraduate program. Also, another common pitfall upon graduation is entering an employment endeavor without adequate research and appropriate data regarding the company. Too often many engineering students graduate and begin working for companies that are not in harmony with their career goals and aspirations. To avoid this from happening to you begin the job search process early in your undergraduate career. It's never too early to begin thinking about the type of company you want to work for. Consult mentors, professors, family, and friends to assist you in this process. Also, explore books and magazines that examine various advantages and disadvantages to working for particular companies. Moreover job fairs provide a wonderful opportunity to explore and inquire about different companies. Following, the aforementioned strategies will put you ahead of your competition and may create a mutually beneficial situation for you and your company.

VII. Qualifications Required

A bachelor's degree in industrial engineering is specifically required for jobs in industrial engineering. The training you receive from your undergraduate institution is typically sufficient to succeed in the field. It is not uncommon for engineering degree holders to work in similarly related engineering fields. Although graduate degrees (M.S. or Ph.D.) in industrial engineering are not required the attainment of such advanced degrees may lead to promotion and higher salary compared to bachelor's degree holders.

(i)nteresting Facts

❶ Engineers typically work a forty-hour week and are employed in the cities and rural areas of every state.

❷ There are approximately 330 colleges and universities that offer bachelor's degrees in engineering.

❸ Access the following website for more information about careers in engineering: *http://www.nspe.org*.

Source: U.S. Department of Labor. *Occupational Outlook Handbook*, 2002–03 ed. Indianapolis, IN: JIST Publishing, 2002.

Take Home Points

- Industrial engineering deals with the design, improvement, and manufacturing of materials in an industry setting
- Starting salaries for persons with a degree in engineering are typically higher than that of graduates from most other disciplines
- In addition to a comprehensive knowledge of engineering and mathematics and related scientific disciplines you also must possess other important skills that will ensure a successful career
- Money can make you happy temporarily, but doing what makes you happy and what you are passionate about will bring you happiness forever
- Begin the job search process early in your undergraduate career
- A bachelor's degree in industrial engineering is specifically required for jobs in industrial engineering

Suggested Reading List

Baine, C. *Is There an Engineer Inside You?: A Comprehensive Guide to Career Decisions in Engineering*. Calhoun, LA: Bonamy Pub., 1998.

Garner, G. *Great Jobs for Engineering Majors*. Columbus, OH: McGraw-Hill, 2002.

Yuzuriha, T. *How to Succeed as an Engineer: A Practical Guide to Enhance Your Career*. Vancouver: J&K Pub., 1998.

GENETICS

> To be who you are and become what you are capable of is the only goal worth living for.
>
> —*Alvin Ailey*

Mrs. Barbara W. Harrison

Certified Genetic Counselor
Department of Pediatrics and Child Health
Howard University

I. Biographical Sketch

In school, I always excelled in mathematics and science, undoubtedly due to the influence of my father, who was a mechanical engineer. During high school, my interest specifically in biology grew, particularly in topics related to how the human body functions, including cell biology, anatomy, and genetics. In addition to my academic studies, my interest in genetics was also peaked by my two cousins, both of whom have sickle cell disease. For most of my childhood, I was oblivious to any complications they were having. With age, however, I noticed that one cousin always seemed thinner than others in the family and the other cousin consistently had problems with her knee.

I enrolled at the University of Maryland, College Park, as an undergraduate biology major. After taking introductory biology courses, I began to gravitate to classes on genetics. I found genetics, the branch of biology that focuses on heredity and variation of biological organisms, to be extremely interesting and challenging. I graduated *cum laude* with my degree in biology with an emphasis in genetics. I accepted an opportunity to work with a teaching professor at the National Institutes of Health during the summer between my senior year and graduate school.

In the meantime, I was searching for a career. I was interested in medicine, yet not motivated to become a physician or nurse. Although having responsibility for saving lives is very noble, it was not one that I was willing to take. In addition, the lifestyle of most physicians I knew was not in line with the type of family life I wanted to have. Another option was research, and most of the classes I took were geared in this direction. I wasn't particularly excited about spending all of my days at a laboratory bench, but I decided to look further into it.

The turning point for me in my career search came while speaking to a professor about his institution's Ph.D. program in genetics. This professor inquired about my research interests, to which I replied that I would like to find a project where I could interact with people as well as work in the laboratory. After giving me a questioning look, he asked if I had ever heard of genetic counseling, a term I had seen in my textbooks, but never had a definitive grasp on. The professor suggested that I meet with one of his colleagues who was a genetic counselor. After a very informative and enlightening meeting, I was on the road to becoming a genetic counselor.

I matriculated in the genetic counseling training program at the University of Pittsburgh. While there I had the opportunity to learn even more about human and medical genetics and had the opportunity to see clients. As part of all accredited genetic counseling programs, clinical rotations at approved sites are required to obtain experience working with clients. Documenting supervised experience with clients is an essential requirement to becoming certified in the field of genetic counseling. Students see clients that are referred for a variety of reasons, giving them a well-rounded experience, and they have the opportunity to work with genetic counselors, geneticists, physicians, and other health professionals, and have the opportunity to network. Students' performance at their clinical rotations is very important to their future advancement in the field; they will look to their supervisors for personal references, and some may try to eventually work at the genetic centers where they rotated. In addition, the learning curve during the time of rotations is very steep; not only are you learning about different genetic syndromes and practical issues of providing health care, but you are also gaining knowledge about the counseling aspect of genetic counseling, including dealing with patients who are making very difficult decisions that often affect not only themselves, but their children, parents, and other family members. Therefore, the clinical rotation experience is one of the most important aspects of a genetic counseling program, and should be approached with great care and seriousness by the student.

At the genetic counseling program I attended, thesis research was required, which I pursued at neighboring Carnegie Mellon University, doing molecular genetics work involving sickle hemoglobin. The goal of my research was to determine the effect of two naturally occurring point mutations on the sickling process of hemoglobin S. The goal of this type of research is to identify amino acid interactions in the sickle hemoglobin polymer that, if disrupted, will decrease the amount of cell sickling that occurs. If such interactions can be identified, it's possible that they can be interrupted through the use of pharmaceutical agents or gene therapy. Of note, not all genetic counseling programs require a thesis project, although most involve conducting research on some level, sometimes in the form of a student project. Many students pursue more counseling-related research, as opposed to the type of research that I conducted.

I greatly enjoyed studying human molecular genetics and genetic counseling on both the undergraduate and graduate levels. During my studies and recently thereafter, the Human Genome Project has brought to the public's attention the importance of genetics on our health. The purpose of the Human Genome Project is to determine the sequence of nucleotides of the genes of the human body. Our knowledge base in genetics, related particularly to unraveling the genome and identifying the roles of various genes, is still growing at an astronomical rate, and we are in the midst of a major revolution in the delivery of health care. It is very exciting being a part of that revolution, not only from the scientific aspect, but also from the human aspect, in trying to determine the effect of this knowledge on the public's health and society at large. The plethora of ethical, legal, and social issues that are now in front of us is challenging, and with the scientific advances occurring at such a quick rate, we must deal with these issues in a way that will preserve our well-being.

After graduating from Pittsburgh in 1995, I accepted a grant position shared by Sinai Hospital in Baltimore and Howard University in Washington, DC. This was a unique opportunity for me, because I was able to see patients and also do community work, particularly in the African American community, geared to educate the public about genetics and to recruit more people of minority background into the field. In addition, my strong feelings of accepting and advocating for those of varying disability began to emerge, something that was first instilled in me during my childhood by my mother, who was a special education teacher. A few years later, I switched to working solely at Howard University and now assume several responsibilities there.

I am a member of the National Society of Genetic Counselors (NSGC) and am co-chair for the Diversity Special Interest Group and part of the membership committee. I have presented research at the annual educational conference of the NSGC and published in the *Journal of Genetic Counseling*, the *Journal of Molecular Biology*, and the *Journals of Registry Management*. Most recently, I was appointed to the secretary's advisory committee on genetics, health, and society, formed under the department of health and human services.

I have received a variety of honors, awards, and scholarships throughout my undergraduate and graduate education. I attended the University of Maryland tuition room and board-free through receipt of the Benjamin Banneker Merit Scholarship. While there, I was accepted into the Phi Kappa Phi Honor Society, Mortar Board National Honor Society, and Golden Key National Honor Society. Likewise, at the University of Pittsburgh, I was the recipient of the K. Leroy Irvis Provost Scholarship which paid for my tuition during my first year of graduate school, my second year was paid for through an NIH Minority Graduate Research Grant Supplement related to the research I undertook during the completion of my thesis project.

II. Description of Job Duties

Most people are not familiar with genetic counseling, but this has been changing over the past decade. As described by the National Society of Genetic Counselors (www.nsgc.org), "... genetic counselors typically work as members of a health care team, providing information and support to families who have members with birth defects or genetic disorders and to families who may be at risk for a variety of inherited conditions. They identify families at risk, investigate the problem present in the family, interpret information about the disorder, analyze inheritance patterns and risks of recurrence, and review available options with the family. Genetic counselors also provide supportive counseling to families, serve as patient advocates and refer individuals and families to community or state support services." Some genetic counselors work in strictly clinical roles, seeing patients and providing these types of services. The number of patients seen by a genetic counselor on a weekly basis varies from center to center, from as little as two or three to as many as thirty. Like myself, about half of genetic counselors work in university medical centers, while others work for private companies that provide genetic services (counseling and testing). I am the genetic counselor for Howard University Hospital and see patients from a variety of referral sources, including prenatal, pediatric, and cancer.

Some genetic counselors work in the academic arena, and may or may not see patients. They can be actively engaged in research activities related to medical genetics and genetic counseling. They may be the principal investigators (PI) of such research or may be part of the research team, recruiting and interacting with patients and families involved in various studies. I am co-PI for a clinical research study which offers biochemical and genetic screening for hereditary hemochromatosis, a type of iron overload disorder. I see individuals who have high iron levels and may have inherited an altered gene that causes them to absorb more iron from their food than necessary. During a typical consultation I explain to them their test results and their risks for future medical problems and discuss possibly having their family members tested.

Some genetic counselors teach undergraduate, medical, genetic counseling, or other students in health professional fields, or work in administrative capacities in directing or assisting with a genetic counseling training program. Even counselors not directly affiliated with an academic institution may supervise genetic counseling students who are in clinical rotations or medical students and residents. Currently, I am co-director of the genetic counseling program at Howard, which involves working with master's level students acquiring their genetic counseling degrees. I teach a course taken during their second year, organize their rotation experiences, supervise students rotating at Howard University Hospital, and work with them on their thesis projects, sometimes as their main thesis advisor. I am also an adjunct professor of the health care ethics program, which teaches a course in health care ethics to medical, nursing, dental, pharmacy, and allied health students.

Genetic counselors can work with a variety of other companies and institutions including: diagnostic laboratories, working as a liaison between physicians and patients, and possibly doing laboratory work themselves; for the pharmaceutical industry, as a genetics education specialist developing educational programs for various business groups; in public health, on local, state, and federal levels, coordinating genetic services and newborn screening, educating physicians and the public about genetics, constructing and maintaining a birth defects registry or other related registry, or participating in the development of public policy; or for Internet companies, as an expert or consultant in writing genetics information for a website, or developing online resources for physicians and the public. Genetic counselors may also work in private practice, negotiating contracts and services with local health care facilities to provide services for patients and families.

Grant writing can be an important aspect of a genetic counselor's position, but varies greatly depending on where he or she is employed. Some counselors are involved in research and apply for grants regularly, although this is sometimes done in conjunction with an established researcher, either in genetics or psychology. There are some grants given by groups like the National Society of Genetic Counselors that are designed for genetic counselors to apply for as principal investigators. Some genetic counselors have positions that are secured by grant funding, even on the clinical level, as we are currently not able to bill directly for our services. As mentioned earlier, I have been co-PI on two grants, one called "Providing Access to and Utilization of Genetic Services among an Indigent African-American Patient Population at Howard University Hospital," which was a service grant, and "Hereditary Hemochromatosis and Iron Overload Study," which is a clinical research grant.

III. Advantages of Career

One of the main aspects that draw people into the field of genetic counseling is a love of science, but also a desire to interact with people. Like myself, many individuals in this field could have pursued more traditional careers like medicine or nursing, but chose genetic counseling because they had a particular interest in genetics. The work of the Human Genome Project and other recent developments in genetics have also incited interest and opened the door to many opportunities for genetic counselors. These developments have expanded the field of medical genetics from a narrow list of relatively rare medical conditions to more common conditions like cancer, diabetes, and Alzheimer's disease. As more genes are being identified and their functions revealed, there are more people in our society who will face decisions regarding genetic testing. This also creates many ethical and moral dilemmas, and genetic counselors are uniquely equipped to sit at the table with ethicists, scientists, and policymakers as decisions are made about legislative issues. With time, other health care professionals are realizing the contribution that genetic counselors can provide and are taking advantage of our clinical and research expertise.

The opportunity to have a variety of duties, as demonstrated by my current position, is certainly an advantage. Every day presents new challenges, which can come from the clinical, research, or educational realm of my position. In addition, I am invited to participate in meetings of experts to develop educational tools about genetics or survey instruments to determine the public's understanding of genetics, make presentations to community groups to provide genetics education, and speak to high school students and undergraduates about pursuing a career in genetics or genetic counseling. Other counselors only see patients, which presents plenty of variety as well. Each patient has unique familial backgrounds and personal qualities that make each experience different.

In most settings, the genetic counselor is able to spend more time with patients than the physician and is able to appreciate the cultural, social, and religious views that patients have that will affect their decision about genetic testing or how they handle the results of such testing. I see this as an advantage, because patients will likely see the counselor as an advocate for their interests, and this can be crucial to patients who are usually in very stressful situations at the times when we are interacting with them. It's always a rewarding feeling when a patient expresses gratitude to you for helping him or her through a difficult time.

IV. Essential Related Skills for Success

Genetic counselors must have the ability to interact with people on a personal level. In my experience, many applicants to genetic counseling programs have high marks in their biology and genetics classes, but have no evidence of people skills. These types of skills can be developed and demonstrated by working on a crisis hotline, maybe involving rape victims or persons contemplating suicide. The ability to talk people through high-stress situations is a crucial skill in genetic counseling. Experience working with people with disabilities, mental or physical, is a way to show that you have a compassionate spirit, something that is almost essential for individuals in this field. This spirit is something that needs to be apparent to clients when they meet you, or they will see you as just another health professional involved in their care, as opposed to someone they can trust and with whom they feel comfortable and safe expressing their true feelings.

Many counseling theories are employed by different genetic counselors as they serve their patients. An underlying theory that has maintained a crucial role in genetic counseling is called "nondirectiveness." This term, though widely used in genetic counseling training, has come under debate regarding its specific meaning. Ideally, it means not steering clients toward making a decision based on your personal beliefs. More specifically, this term means the counselor must share information with the client, and then assist the client in various decision-making processes by listening, asking questions, and respecting the client's social and cultural background. The role of the genetic counselor varies greatly depending on the client, because clients have varying needs and expectations. Carl Roger's client-centered therapy and more recent literature written by Seymour Kessler, Heinz Kohut's self-psychology theory, the Stone Center's Self in Relation Theory, and various family systems theories have been employed by genetic counselors; major concepts within these include doing a psychosocial assessment of the patient, showing unconditional positive regard for patients, listening to the patient's story, promoting competence and autonomy, reframing, expressing empathy, and exuding overall genuineness and care for the client. These theories and concepts, when applied appropriately, usually create a comfortable and safe atmosphere for the client to express intimate feelings. Some clients don't know what to expect when they see a genetic counselor and remain reserved and quiet throughout the session. However, most clients, once convinced that the genetic counselor is there to help and not judge or force a particular action, are able to see the counselor as a confidante and rely on them to help them come to a decision.

V. Additional Advice

Although most applicants to genetic counseling programs have a background in science, I have also seen those with a psychology, social work, and social science background pursue genetic counseling. These individuals bring special skills and experiences with them that are particularly suited for this field, but a certain level of competency in science is required. Likewise, it is helpful for any student who is interested in genetic counseling to take at least one psychology or counseling course during their undergraduate career.

There are currently about thirty master's level genetic counseling programs in the United States, Canada, and abroad. Different programs have different prerequisites and emphasize different aspects of the field, be it more scientific, more psychosocial, or more geared toward research. It is important that you do your homework to find the program that will fit your needs.

Finding a mentor or mentors who are knowledgeable about genetics and the genetic counseling field is extremely helpful in successfully applying to and completing a genetic counseling program. These individuals can guide you through determining an appropriate career path and assisting in the process of obtaining your degree. Although finding a mentor in the genetic counseling field is best, it may not be possible. Therefore, you can find guidance through a researcher in human genetics and maybe a social worker or psychologist.

Properly researching the field of genetic counseling is also critical before deciding to apply to a program. It is most ideal to secure some kind of volunteer or paid experience in a center that provides genetic counseling or a related service. At the very least, you should seek out a genetic counselor in your area to visit and possibly shadow for a day or two to ensure your genuine interest in the field. Genetic counseling programs invest a large amount of time and effort into each student, and are therefore careful in choosing individuals they feel have a good understanding of the career and will contribute significantly to the field upon graduation. This is also in your best interest, as you don't want to invest time and money into training for a career that you eventually do not pursue.

As previously mentioned, experiences on crisis hotlines or with people who are disabled or disadvantaged, or even working with tutoring or teaching programs provide excellent opportunities to prepare for a career in genetic counseling. You may pursue this over the summer or during the school year through campus- or community-based organizations. Such groups are almost always looking for volunteers, usually provide training, and have a variety of options for involvement.

Regarding graduate education in general, it is always worth the effort to try to secure funding to assist in the costs of higher education. There are many groups, private and public, that offer scholarships and grants for students, sometimes based on merit, on need, or on the program pursued. One thousand dollars here and there can add up and at least assist in buying books or other supplies, as well as chipping away at the cost of tuition. With the Internet at your fingertips, searching for such opportunities has never been easier.

VI. Pitfalls to Avoid

In general, researching a career is the most important step in one's career search. There are many rewarding opportunities in medicine and allied health, including dentistry, physical or occupational therapy, pharmacy, nursing, or becoming a physician's assistant. It is important that you research these options before jumping in with both feet, as they all require a certain level of commitment. The biggest pitfall for anyone is pursuing something that you end up not being interested in by the end of the program. Many times this can be avoided by proper preparation.

VII. Qualifications Required

To become a practicing genetic counselor, you must first obtain a bachelor's and a degree from an accredited master's level genetic counseling program. Accreditation is granted by the American Board of Genetic Counseling (ABGC). Different programs have different prerequisites, from required coursework to community service or experience outside the academic realm. Programs range from a year and a half to two and a half years. Some require research in the form of a thesis or student project. Some programs emphasize the scientific aspects of medical genetics while others concentrate on psychosocial issues.

Certification for genetic counselors is accomplished by passing an examination administered by the ABGC. Genetic counselors who are not certified can currently practice and see patients, although it may be required at some centers that they be supervised by a certified genetic counselor. However, as there are movements toward licensure in many states and recognition by insurance companies of genetic counselors as billable health care providers, certification will become a professional necessity.

(i)nteresting Facts

❶ Obtaining biological information, determining risks involved, offering recommendations, and employing counseling strategies are typical activities for genetic counselors.

❷ More information about genetic counseling can be acquired from: *http://www.abgc.net.*

❸ Increased employment opportunities in genetic counseling will be made available in future years.

Source: U.S. Department of Labor. *Occupational Outlook Handbook,* 2002–03 ed. Indianapolis, IN: JIST Publishing, 2002.

Take Home Points

- Genetic counselors can work as members of a health care team. They provide information and support to families who have members with birth defects or genetic disorders and to families who may be at risk for a variety of inherited conditions
- Genetic counselors may work in research, public health, or administration
- Not only is academic knowledge important, but experience in counseling persons in high stress situations is essential
- Be sure to select a mentor who is knowledgeable about genetic counseling. Work at a genetic center or shadow a genetic counselor to gain experience
- Research the field of genetic counseling to ensure that you will be satisfied with it as a career
- Completing a master's degree in genetic counseling is required to practice. Attaining your certification is becoming increasingly important for longevity in the field

Suggested Reading List

Baker, D., Schuette, J., & Uhlmann, W., eds. *A Guide to Genetic Counseling*. New York: Wiley-Liss, Inc., 1998.

Kelly, T. *Clinical Genetics and Genetic Counseling*. Chicago, IL: Year Book Medical Publishers, 1980.

Reed, S. *Counseling in Medical Genetics*. 2nd ed. Philadelphia, PA: W. B. Saunders, 1963.

GEOLOGY

> All my life I've wanted to be somebody. But I see now I should have been more specific.
>
> —*Jane Wagner*

Dr. Carol J. Zimmerman

Geophysical Associate
Geophysical Applications Group
ExxonMobil Exploration Company

I. Biographical Sketch

I came to my profession, a geophysical applications specialist for a major oil company, in a unique way. Geophysics is the application of physics, geology, and mathematics to help understand what lies beneath the surface of the earth. Early on, I decided to pursue a career in the sciences because I had a talent for mathematics. Mathematics provided me with logic, structure, and answers.

I won many schoolwide contests in mathematics in junior high school. A particular teacher, Mr. Burich, was my mentor for three years. In high school I took calculus with another mentor, Mr. Harrison. These mentors were very important, because they encouraged me, and worked with me as a coach would work with an athletic team. As a high school junior, I won the Rensselaer Science and Math Award, and a special scholarship for my high SAT scores. Contributions from these two sources, and a small student loan, paid for my undergraduate education at Rensselaer Polytechnic Institute. There, I majored in the most basic of sciences, physics, a discipline where I could apply my mathematical skills.

I found physics fascinating because I learned that it is really the basis of all the other sciences. It is an old science, with an interesting history. The people who made the great strides in physics are incredible personalities as well as being incredibly brilliant. Even the physics professors of today are unusual characters. One of my physics professors, Robert Resnick, wrote a basic textbook in physics, which is still used today. It is a science, which is always moving forward to discover new things through mathematics and experimentation, but the basic premises and methodology stay the same.

Even when a great discovery alters the way that the world is viewed, such as Einstein's Theory of Relativity, the discovery sits side by side with the old, which in this case is Newtonian mechanics. The old hypothesis explains certain situations, whereas the new one pertains to other, newly discovered situations. The new hypothesis can be reduced to the old one mathematically to fit the circumstances. So both theories are correct. Therefore, basic premises of physics are never out of date, just as in mathematics. They just build upon one another.

I managed to secure a teaching assistantship at Rensselaer for one semester, where I taught two physics classes, one to the honors class and the other to a remedial class. The honors group was always challenging, whereas the remedial group needed a lot of extra help from me. This taught me how to communicate the same material to people of different abilities.

After I graduated from Rensselaer, I matriculated at the University of Wisconsin, but I was unable to secure a research or teaching assistantship in physics to pay for my college expenses. A kind professor, Dr. Robert Moore, took me under his wing, because he knew how badly I wanted to continue with my education. He offered me a research assistantship, but I would have to change my major to oceanography and limnology (the study of lakes). This sounded very interesting to me, because I could apply scientific principles that I learned in physics to real-life situations.

The major that I had taken under the oceanography program was in geology, the study of the earth. I did geological exploration for noble metals, gold, and platinum in estuaries of the Bering Sea in Alaska as my research project for my master's degree. To do this, we analyzed surface rock samples that we collected.

Later, I went on to a research assistantship with Dr. Moore's best friend, Dr. Robert Meyer, in geophysical oceanography. Geophysics leans more towards the physics and mathematical side of earth studies that tied in very well with my undergraduate work. I had both research and teaching assistantships to finish up my Ph.D. Also, I received a scholarship from the Society of Exploration Geophysicists, a professional society, which I still belong to today. My research was still in exploration, in offshore remote detection of sand and gravel deposits in Lake Michigan. By this, I discovered a previously unknown southern extension of a glacial till. I wrote scientific articles and spoke at professional society meetings while still in school, which greatly improved my self-confidence and determination to stay in geophysics.

After my Ph.D., I decided that I really wanted to do research. I was offered many exploration positions with oil companies, because of my exploration experience. I decided that I preferred to join a research organization, where I could develop new geophysics experiments. This led me to the Exxon Production Research Company in Houston, Texas. I remained there for eleven years. While there, I was the author of three inventions that were patented. As a result of one of my patents, I transferred to the exploration division of Exxon U.S.A. in order to apply my invention to exploration problems. Within the year, Exxon Exploration Company was formed. I remain here today, eleven years later, although it is now called ExxonMobil Exploration after a merger.

My current field still is in exploration. Here, I work in a geophysical applications group, which provides state-of-the-art high technology services to exploration geologists who develop the prospects, i.e., places that we want to drill for oil or gas.

II. Description of Job Duties

Exploration is not a field like mathematics where an answer can be calculated before a well is drilled. It is risky business. When a prospect is developed, there are many unknowns. My group tries to help define the geophysical risks involved before drilling a prospect and to help provide the error bars on the uncertainties.

Seismic data, our primary tool, is acquired by making a loud noise on the surface of the earth or in the water and recording the echoes from the subsurface of the earth with many receivers. After extensive processing of these data, the echoes produce a three-dimensional picture of what lies beneath the surface of the earth. Although it may not look like a picture to someone who is not acquainted with these types of data, a good seismic interpreter can see the picture and map the geologic formations. Seismic interpreters generally are geologists. Geologists study the composition, form, and structure of the earth. These persons vigorously study rock formations and can tell by their shapes what might be below the surface. Geologists can also pontificate on how the rocks were originally deposited, ages ago, by comparing them to more modern rocks they have learned about. So, the use of seismic data is one way of 'seeing' below the surface of the earth to test where there may be hydrocarbons, oil and gas deposits, for which we want to drill.

Sometimes gravity data also is incorporated. These data measure the density of the formations below the surface of the earth by measuring the pull of gravity upon them. Oil and gas deposits typically are less dense than water that is contained in the rocks, and so gravity data may help us to find these less dense areas. When these data are combined with the seismic picture, a pool of hydrocarbons may be found.

Data from wells already drilled, whether or not they contain hydrocarbons, also help us to evaluate new exploration areas, as well as to help us develop deposits already discovered. Cuttings from the rocks encountered by the drilling tool can be examined by another type of geologist, the formation evaluation specialist.

Logging tools also are available which can be put down the well to tell us what kind of rocks we have encountered, what kind of fluids are in them, and exactly how deep they are. These data also must be interpreted before sense can be made of them. The formation evaluation specialists also examine these data. I handle the geophysical aspects of the well data, which means that I correlate the well data to the seismic data by modeling, again incorporating mathematics. Teams of people from other specialties work with the general geologists and with my geophysical applications team to help mitigate the risks in drilling a prospect. Still, there are more unknowns than we would like to have. That is where research comes in with new tools and new mathematical equations that we can use in our exploration work.

My daily activities are mainly spent working with a two screen, high-powered workstation. This is a computer, which is able to compute both locally, and also network with other computers. Advanced computing power is necessary because of the huge volume of data with which we work and because of the massive calculations that are needed.

Generally, exploration geologists interpret seismic data qualitatively, but my work is more mathematical. I do things with the seismic data to help check the robustness of the geologist's interpretation. One of the things I do is work with the geophysical data acquisitions group and the data processors to check the quality of the seismic data. Another thing I do is to help the interpreter with predictions to the prospect by matching data we learned from nearby wells to the seismic data. I also model the seismic data to make the interpretation easier for the geologist. I also help to provide better 3-D pictures of the seismic data so that the prospects can be visualized from different perspectives.

III. Advantages of Career

The biggest reward that I get from my career is when a successful well is drilled and I have played a major role in helping management to make the decision to drill that well. I have attained many awards for helping in a major way to drill successful wells. These are trophies consisting of clear Lucite blocks with a sample of oil from each well, the well name, water depth, and the names of other oil company partners that participated in drilling the wells. Of course, there is much celebration in the office when a successful well is drilled.

Since we do not deal with the public, this is a very informal work environment. Casual business dress is the norm, and everybody calls each other by their first names, even when talking to the president of the company.

Other rewards consist of continuous learning and varied work experiences both in the computer work and in working with different groups in the company. The work is challenging.

There is also always the opportunity to travel or to live in other parts of the world. I find that even if I do not travel, there is great diversity in the workforce. I am always surrounded by interesting people, and am always learning something new about other cultures of the world. My work group is always changing personnel, giving a fresh perspective on the particular problems of each geographical area. So even though we are working for a large company, it really is like living in a small town. After one has been here long enough, he or she knows many of the people who work for the company, because of the opportunity to have worked with them at one time or another. I also take pride in knowing that I am helping to meet the energy demands of this country and the world.

IV. Essential Related Skills for Success

Just understanding the computer programs and the science behind them is not enough to become successful in the work here. Interpersonal skills always are a plus. Everyone must get along with one another and be able to work in a team environment. Although it is fine to have professional disagreements, they are presented in a way that is not demeaning to anyone else. Respect for others is a must.

Next to respect, good communication skills are a must. These can entail dealing one-on-one with another colleague or other employee, dealing with a team, or presenting a complex idea to middle or upper management. Good presentation skills are necessary. Presentations must be given in a way that is understandable to someone who is not in your field of expertise. One cannot be shy in giving a talk to many people, or at a professional society meeting. The talk must be well organized and geared to the background of each particular audience. For example, a talk given to the geophysical applications group can be more technical than one given to management. Good writing skills are equally important. Written reports are required for documentation. These must be well organized and must be understandable to a wide audience.

Project management skills also are important. A project is expected to be completed within the time frame allotted, even if that is a short time frame. One must know when to do a detailed analysis and when to hit the high points in order to finish on time. The objectives of the project must always be understood and met. One cannot go off on a tangent, unless that tangent leads to a quicker, less costly, or better answer.

Good mentoring skills may also be needed to help temporary or new employees come up to speed and to further assist current employees. Fortunately, the company provides additional learning opportunities to help employees develop the above skills. Also, training is available to help employees with stress management, and other important personal skills.

V. Additional Advice

If you are considering a career with an oil company, you may wish to join the Society of Exploration Geophysicists, the Society of Exploration Geologists, the Society of Petroleum Engineers, or any other related organizations. This will give you a feel for what the work entails. I joined two professional societies, the Society of Exploration Geophysicists and the American Geophysical Union, while I was still a graduate student. From these organizations you will have the opportunity to obtain informative publications you can read for further information about the field. Each of these organizations has an annual meeting where technical papers are presented, and where vendors have booths to show off their latest technical developments. I participated in the annual meetings by giving presentations to large audiences. This greatly improved my self-confidence and speaking skills. Most universities will pay for your expenses to attend these meetings if you are giving a presentation, because it is good networking for the university professors, who may work part time for or receive grants from industry for research projects.

Apart from giving a talk yourself, by attending meetings of professional societies, you are able to listen to talks given by others, which improves your knowledge of the field. The meetings introduce you to some of the problems that have been solved, as well as those that still need to be solved, within the industry. The meetings also give you an idea of what skills you may need to help solve these problems.

Don't think that once you join a large company like ExxonMobil that you will necessarily stay in the same town or even stay in your present field. You have to help guide the direction of your career by use of good mentors, by communicating your training desires, and by having a good relationship with your supervisor and other superiors. Larger companies are more economically stable and may offer better benefits than smaller companies.

The major oil companies usually offer a summer hire program for students so that you can get a feel for the job. After school, your success is based upon your performance on the job, not on your education level.

If you choose to work in the oil business for a smaller company, or for a vendor, your chosen field may be easier for you to control, and you may make a larger salary, but your position is a bit more tenuous. You may not be able to make a career with the same company. Mergers happen. Vendors go belly up. You may find yourself jumping from company to company in order to find a position that is suitable to your skills. Smaller companies and vendors do not offer the training opportunities that a large company does, and they want to hire people who already have the required skills for the particular job opening.

Having a good mentor while you are in school can be very instrumental. I was fortunate to work with several professors who worked for oil companies prior to their teaching at the university. They gave me a good idea of what working for an oil company would be like, and how to prepare myself for it. You may also meet good mentors who are working in the industry at the professional society meetings.

VI. Pitfalls to Avoid

The major obstacle you may encounter in obtaining a position in the field of oil exploration would be the economy. I soon found out that a career with an oil company is subject to the ups and downs in the economy. Although we are compensated well, our jobs are predicated on these ups and downs. Only the larger, stronger companies survive a serious fall in oil prices or the economy.

I feel very lucky to have made a career here. Oil companies can go for many years without hiring. You must graduate at a time when there are jobs available. This is hard to predict, and I can't really offer a good solution to this problem, except to obtain a job in a related field until times get better. This is also where a Ph.D. might come in handy, because one can take a graduate assistantship or teach to buy time until the oil companies start hiring again. If you do not have a Ph.D., you may want to work for a minerals exploration firm, which would further prepare you for oil exploration.

I don't think I could have prepared better for my present career, or hit the market at any better time than I did. I was very lucky. I had a basic science under my belt, which I could fall back on, had worked and studied in both management and in exploration fields, and had a very broad scientific background, which integrated many different sciences. When I graduated from the university with my Ph.D., the oil business was booming.

VII. Qualifications Required

A Ph.D. is not required for an oil company exploration career. A master's degree helps a lot, but there also are positions for persons with a bachelor's degree. There also are positions for technical support for graduates of technical schools and two-year colleges. It just depends upon how far you wish to go with your career. It takes at least a bachelor's degree to make it into the professional levels or into management.

(i)nteresting Facts

❶ More information on career opportunities in geophysics can be obtained from: *http://www.agu.org*.

❷ Geoscientists often work in remote field locations.

❸ Petroleum, mineral, and mining industries typically offer higher salaries than other industrial employment opportunities.

Source: U.S. Department of Labor. *Occupational Outlook Handbook*, 2002–03 ed. Indianapolis, IN: JIST Publishing, 2002.

Take Home Points

- Geologists study the composition, form, and structure of the earth
- Advantages of a career in geology include a comfortable living with great benefits
- Essential related skills for success are presentation skills, writing skills, interpersonal skills, and the ability to respect others and work as a team
- Once you decide on your career aspirations it is your responsibility to get the educational training needed to achieve your goals
- A main pitfall to avoid is assuming that a position for you is available when you are ready and that it is secure once you are there
- Qualifications required for oil exploration range from two-year technical degrees to a Ph.D. A bachelor's or master's degree in geology, physics, oceanography, or chemistry is usually a minimal educational requirement for entry-level positions

Suggested Reading List

Camenson, B. *Great Jobs for Geology Majors*. Lincolnwood, IL: VGM Career Horizons, 2000.

Rose, P. *Guiding Your Career as a Professional Geologist*. Tulsa, OK: Division of Professional Affairs American Association of Petroleum Geologists, 1994.

Rossbacher, L. *Career Opportunities in Geology and the Earth Sciences*. New York: Arco, 1983.

MEDICAL TECHNOLOGY

> He who learns but does not think is lost; he who thinks but does not learn is in danger.
>
> —*Confucius*

Mrs. Karen M. Kiser

Professor
Clinical Laboratory Technology and Phlebotomy
St. Louis Community College at Forest Park

I. Biographical Sketch

I have always had an interest in science. My love for science can be traced back to my childhood. While my twin sister preferred to play with dolls I had an inclination to study the living world using my microscope and chemistry set. I loved performing chemistry experiments and examining the living and non-living world with my microscope. I decided to become a medical technologist in the seventh grade. It came after reading a book in which the main character was a medical technologist.

Medical technology involves many activities such as developing diagnostic protocols, evaluating test results, performing laboratory tests, utilizing experimental instruments, and having a thorough understanding of the scientific concepts underlying the tests performed in the laboratory.

After graduating from high school I decided to attend Southeast Missouri State University (SEMO). While matriculating at SEMO I participated in a summer employment endeavor that afforded me the opportunity to work with laboratory professionals. It was during this endeavor that I confirmed that I had made the right choice to pursue a career in medical technology. I felt right at home in the clinical laboratory. I loved the work and enjoyed the experience. During my junior year of college, I applied to three hospital-based medical technology professional education programs. I attended the Jewish Hospital School of Medical Technology clinical laboratory science program during my senior year. The internship program lasted for fifty weeks and involved intensive training in diverse clinical laboratory settings as well as pertinent lectures on important clinical issues.

Before I graduated from SEMO I was offered four jobs in medical technology related-fields. After attaining my certificate in clinical laboratory science and graduating with a Bachelor's degree in Medical technology from SEMO, I accepted a job offer to work at Jewish Hospital in their microbiology lab. Later I joined the staff of the medical laboratory technician program at St. Louis Community College, Forest Park. I also returned to school and completed a master of arts in health care education at Central Michigan University.

During my tenure at St. Louis Community College, I have been interviewed by a local radio station during National Medical Laboratory Week, been honored as the Teacher of the Year, and published a few articles in the Journal of Clinical Laboratory Sciences and Advance for Medical Laboratory Professionals. I also received the Governor's Award for Teaching Excellence and the Emerson Electric Award for Excellence in Teaching. I am currently a member of the American Society for Microbiology, American Society for Pathology, and the National Education Association.

II. Description of Job Duties

Medical technologists are primarily responsible for the collection and analysis of patient specimens thereby assisting physicians in the diagnosis, treatment, and prevention of disease. Teaching, quality assurance, and inventory control are also keys aspects of a career in medical technology.

The Core Lab

The core lab typically is equipped to handle the analysis of most body fluids and tissues. Sample analysis is performed by many sophisticated instruments. Instruments are not infallible, however, and therefore pre-analytical and post-analytical processes must be conducted. The job of the medical technologist is to constantly monitor the results to determine that they make sense prior to reporting them. Alerting physicians to potentially life-threatening results is also part of the job. The core lab may consist of several laboratories in which specific tests are performed, they are: hematology, coagulation, urinalysis, chemistry, immunology, blood bank, and microbiology. These individual areas will now be described in detail.

Hematology

The hematology laboratory analyzes blood and the blood cell forming organs, such as bone marrow, for abnormalities. An abnormality may indicate an anemia, infection, or leukemia. Tests are also done to monitor therapy, after diagnosis of a hematologic disease. Although instruments actually perform the cell counts now, in many instances they only flag a blood sample as abnormal. The medical technologist must then examine the blood smear to determine the exact nature of the abnormality.

Coagulation

The coagulation laboratory analyzes specimens in order to detect clotting or bleeding disorders and monitor anticoagulant therapy. An abnormal result may indicate hemophilia or anticoagulant over or under medication.

Urinalysis

The urinalysis laboratory evaluates urine specimens for chemical changes or the presence of formed elements that may indicate a problem in the urinary system.

Chemistry

The chemistry laboratory analyzes specimens for various chemicals. These chemicals may exist at normal levels or be elevated in disease. Diabetes is an example. The compound measured is glucose. A high result indicates diabetes. Other chemicals are released only upon damage to tissue. An example of this would be enzymes found in the heart. They are elevated when a patient has had a myocardial infarction or heart attack. Elevated chemistry results may indicate a high risk for disease development. High cholesterol levels, for instance, may indicate an increased risk for developing cardiovascular disease.

Immunology

The immunology laboratory evaluates a patient's immune system. Abnormal results may indicate an infectious disease or autoimmune disease.

Blood Bank

The blood bank laboratory is responsible for typing and cross-matching donor blood and other blood products, for compatibility prior to infusion into a patient. Blood is also screened for contagious diseases. Timely administration of blood to a patient could mean the difference between life and death. Certain patient conditions may also affect the blood specimen giving aberrant results in any of the above departments. The medical technologist must be able to detect those instances and correct the results.

Microbiology

The microbiology laboratory processes patient specimens in order to detect potential agents of bioterrorism or infectious disease. Because many of these specimens also contain commensal organisms or normal flora, the technologist must be able to distinguish between benign and pathogenic microorganisms. Susceptibility testing is also performed on all significant isolates whose response to antibiotics is unpredictable. Medical labs can also be divided into more specialized areas to detect specific microorganisms, cancers, auto-immune and genetic diseases.

Specialized Laboratory Areas

Molecular Testing

Larger hospitals or reference laboratories may centralize their molecular testing on-site, rather than performing the tests in a specific laboratory department. It involves testing for infectious diseases such as hepatitis, HIV, tuberculosis, chlamydia, or gonorrhea. Viral loads or detection of a particular genetic mutation may also be performed to detect resistance to treatment. Furthermore, genetic mutations characteristic of certain cancers or inherited diseases may be determined. This is a growing area of the medical technology laboratory as knowledge gained from the Human Genome Project is put to practical use.

Flow Cytometry

Considered a branch of immunology, flow cytometry is able to evaluate subpopulation of cells using antibodies tagged with fluorescent molecules. The information is used for diagnosis and monitoring treatment of cancers or viral diseases.

Virology

Some viral disease can only be diagnosed by growing the viruses. Viruses cannot be grown on artificial media. They are grown in living cells. The virology lab is responsible for generating growth of viruses in the appropriate cell line. Once growth is detected, identification is based on their effect on the cell line and immunology techniques.

III. Advantages of Career

There are many advantages of this career. Foremost is the opportunity to make a significant contribution to society by being a member of the health care team. Recent data indicates that a significant percentage of medical decisions are based on laboratory data. The "unseen profession" plays a vital role in the delivery of quality health care.

Variety in the job setting is another advantage. You can choose between more manual work and automated work. You can choose to work in a large hospital or reference lab, which may allow you to concentrate on a single discipline. Smaller settings, like small hospitals, clinics, or doctor's offices may allow you to function in many disciplines as a generalist.

As an experienced medical technologist, you may choose to move into an ancillary career. Medical technologists have been hired by computer companies to work on their laboratory information systems. They have also worked in industry providing customer service or sales and quality assurance. Supervision or management, education, forensics, and research are also career options for those interested in medical technology. This major is also good preparation for medical school.

IV. Essential Related Skills for Success

Clinical laboratories are highly computerized; therefore, computer skills are vital. A medical technologist is expected to participate on hospital and laboratory committees. Good communication, interpersonal skills, and the ability to work in groups are essential skills. Curiosity and attention to detail should also be developed.

The medical laboratory can be a hectic environment to work in. Multiple samples may arrive at the same time. Some of these may need to be processed quickly because of a life-threatening patient emergency or the presence of labile substances or microorganisms. Skills such as problem solving, stress management, organization, and the ability to make decisions are vital. Problem-solving skills are needed to troubleshoot specimen and instrument problems that occur.

In addition to science courses, you should appreciate the contribution of general education courses to your future success. Select courses that will help you to develop general thinking, problem-solving skills, and interpersonal skills, such as sociology, psychology, and logic. Public speaking and English courses can also help you develop both written and verbal communication skills. Physical education and courses in the arts may provide you with some tools to handle stress.

V. Additional Advice

Medical technology is a rigorous and challenging curriculum. A less committed individual may give up. Make sure that you enter in a medical technology program informed about the actual work environment. Seek opportunities to volunteer or work as a lab assistant in a hospital laboratory as opportunity permits in order to gain the necessary experience you need to make an informed decision regarding your career.

It turns out that this ability to persevere in the face of difficulty is a good quality to have as a medical technologist. On those days when instruments malfunction, the computer goes down, or the results don't turn out the first time, a good laboratory technologist tries again and works until quality results can be reported. I believe three things have contributed to my success in this field: a terrific internship program, individuals who set positive examples as mentors, and personal goals.

I spent my senior year of college in a hospital-based medical technology program. This unpaid internship program provided a mix of practical real-world experience in the clinical laboratory along with lectures covering medical technology theory. I had the opportunity to get hands-on experience collecting and processing patient specimens, operating instruments, and interacting with members of the health care team. I also learned why the tests are performed, the physiological basis underlying each test, and the principles behind the test systems. I received a number of job offers prior to completion of the program. The internship was a valuable experience and clearly provided a foundation for my success.

During the internship I was taught by individuals who set wonderful examples for me of how I should conduct myself as a medical technologist. This included the microbiology supervisor who would get so excited over the discovery of a parasite in a patient sample that she would go through the lab asking people to come look at it. I watched many of my instructors stay cool and calm during crisis, and persevere until the work was done. One of my teachers encouraged me to pursue education as an alternate career path and supported me when I had doubts.

Personal goal setting has also played an important role in my career. Setting goals gave me a sense of focus and direction and certainly contributed to my success.

VI. Pitfalls to Avoid

I found the science and math courses in college to be quite demanding. The laboratory portion was always easy for me but the theory was tough. I wish I had taken more math and science in high school. I had some initial difficulty in medical technology school meeting their learning requirements. I was expected to achieve an eighty percent on all written examinations. Concept understanding, knowledge of details, and problem solving skills were assessed on each exam. I discovered that I didn't know how to study properly to meet this standard. I learned to review the old information every day prior to concentrating on the new. I learned to spread my study time out over the week. I learned to test myself to check on my knowledge and understanding. I used time management techniques to schedule appropriate study time. By changing my approach to learning, I not only succeeded in the medical technology program but in other aspects of life.

One can get too comfortable in their job or resist change and miss opportunities for advancement. The one constant in life is change. Be open to new experiences and be willing to fail. Failure is not a bad thing if you learn from your mistakes and keep trying. Personal growth is enhanced by analysis of mistakes. Don't be afraid to try new things.

VII. Qualifications Required

Positions available in a clinical laboratory include laboratory assistant, medical laboratory technician, and medical technologist. A medical technologist must complete a formal, structured educational program in medical technology and have a bachelor's degree.

The formal curriculum usually consists of chemistry courses including qualitative chemistry, quantitative chemistry, and organic chemistry. Biology courses include immunology, microbiology, and anatomy and physiology. Math includes algebra and statistics. Courses in each laboratory discipline are also taken. Medical technology theory-based courses are applied during a clinical rotation in each laboratory department.

Most employers will require the applicants to be registry eligible. In addition to the above, work experience credit may be granted in lieu of a formal education program. National certification examinations are strongly encouraged and may be required for licensure in some states. Completion of an accredited laboratory education program with the appropriate college credit or degree, will qualify an individual for this exam.

(i)nteresting Facts

❶ Medical technologists are found in many different professional settings such as clinics, industry, hospital laboratories, and academic institutions.

❷ Due to increasing technological advances the primary work of medical technologists will become more analytical.

❸ A high percentage of positions are currently available in the field of medical technology.

Source: U.S. Department of Labor. *Occupational Outlook Handbook*, 2002–03 ed. Indianapolis, IN: JIST Publishing, 2002.

Take Home Points

- Medical technologists are primarily responsible for the collection and analysis of patient specimens thereby assisting physicians in the diagnosis, treatment, and prevention of disease
- Medical technologists are an important member of the health care team
- Computer, communication, interpersonal skills, and organizational skills are important for the medical technologist
- Internships can provide practical work experience and should be explored to gain more information on the field of medical technology
- Embrace change and be open to new opportunities
- A bachelor's degree and national certification is required for a career as a medical technologist. Check with your state for specific requirements for licensure

Suggested Reading List

Damp, D. *Health Care Job Explosion!: High Growth Health Care Careers and Job Locator*. Moon Township, PA: Bookhaven Press, LLC, 2001.

Earle, V. *Your Career in Medical Technology: Arco's Career Guidance Series*. New York: Arco, 1981.

Karni, K., Caskey, C., & Gerasimo, L. *Opportunities in Clinical Laboratory Science Careers*. New York, NY: McGraw-Hill Professional, 2002.

MICROBIOLOGY

> Chance favors only the prepared mind.
>
> —*Louis Pasteur*

Dr. Peter H. Gilligan

Clinical Microbiologist
Clinical Microbiology-Immunology Laboratories and Phlebotomy Services
University of North Carolina Hospitals

I. Biographical Sketch

If you had asked me when I was beginning college what career path I would take, being a clinical microbiologist was not in the realm of possibilities. Having grown up as the son of a small town general internist in the Berkshire Hills, a rural region in western Massachusetts, I had the idea that being a physician was the only realistic career choice for a young person interested in a career in the medical sciences.

As a child, I developed an early interest in science. Three events, even today, stand out as being influential in my decision to pursue a career in science and unwittingly a career in microbiology. When I was around ten years of age, my parents gave me a microscope. Although it was a cheap plastic one, its lenses were powerful enough to reveal an unseen world to me. I remember putting the microscope on my dresser and turning on both lights so I could get enough light reflected into the microscope lenses to see whatever it was I was interested in examining such as hairs, dust, scrapings from my teeth, and certainly many other more disgusting secretions that need not be detailed here. I spent many happy hours in the contemplation of this unseen world.

Soon there after, I picked up a cheap paperback book called, *The Microbe Hunters.* I remember lying on my bed instead of being outside in the glorious Berkshire summer weather, being spirited away to a world that was full of excitement and adventure. For many of my generation who became microbiologists, this book was an inspirational introduction to the world of medical microbiology. The work of many scientists who built the foundation of this science is detailed in this book. Here I had my introduction to my great hero, Louis Pasteur.

The third event in my life was a science fair project that I did when I was a sophomore in high school. I did the science fair project for two reasons. First, I am sure my mother "strongly" encouraged it. Just as importantly, I was probably bored by what I was being taught in science at the time, and this gave me an opportunity to do something creative. Science fairs were very different from those that go on today in high schools where the level of scientific experimentation is very sophisticated with the students often being assisted by professional scientists. My science project was a truly independent affair and one that would now be done as an elementary school science project. I looked at the role of tooth brushing in changing the microflora of the teeth. The most important lesson that I learned from this experience is how exciting doing experiments can be.

In college, I was drawn into the maelstrom of being "pre-med." I was not prepared for it either intellectually but more important emotionally. As a result my academic performance was mediocre and I found myself drifting, not sure what I would do with my life now that it was clear that plan A, being a physician, was quickly fading as an option. I didn't have a plan B but one was soon to present itself when I took introductory microbiology in my junior year. To be honest, at that point in my life I was going through the motions academically. I liked college but I hadn't found anything that really fired my imagination. However, Dr. Lingappa's course changed all that. Here I was connected back with my childhood passion for the world unseen, microbiology. I loved everything about the course, the lectures, the textbook, and especially the labs. This great experience led me to do independent study with Dr. Lingappa where I worked in his laboratory a couple of afternoons a week learning about dermatophytes, fungi that infects nails, hair, and skin.

It was now time to make a decision about my future. Foolishly and unrealistically I applied to medical school and not surprisingly was quickly rejected. I needed to figure out plan B quickly as it was December and I was to graduate in six months. I decided to apply to graduate school in microbiology. Six applications were rejected. I subsequently learned that admission to the microbiology programs to which I applied was more competitive than medical school. I finally got accepted to the microbiology program at the University of Kansas. However, I was to be on academic probation because of my less than stellar college record and if I got below a B in any course my first year, I would be dismissed. I loved the courses and performed well in the classroom in graduate school even in my nemesis, advanced organic chemistry. However, the main focus of graduate education in the sciences is to do research. I was fortunate that I had a research mentor who was perfect for me. Don Robertson was a young assistant professor at KU in whose laboratory I chose to work. Over the next four years, he taught me a great deal about being a scientist and also how to be a mentor to young people. Dr. Robertson or "Doc" as we affectionately called him was just beginning on a new avenue of research, studying the toxins produced by an organism called enterotoxigenic *Escherichia coli.* This organism is an important cause of fatal diarrheal disease in children in the developing world. Working on this project taught me two things. First, science was very exciting and that I loved doing experiments. I felt some kindredship with my great hero Louis Pasteur. Secondly, I learned that basic science research is very difficult and challenging.

I was having a particularly hard time seeing how my research had any relevance to helping others in the near term. I was nearing the end of my graduate school education and I needed to figure out where plan B was leading me. After careful consideration I decided to participate in a training program in medical and public health microbiology. This would give me the opportunity to work in a hospital microbiology lab that had an immediate impact on the care of patients. This really appealed to me because I knew I wanted to do something in the field of medicine but there was much about being a physician which did not interest me but I knew that the field of infectious diseases did. I also knew that I did not want to spend my time doing research whose benefit was in the long term. I wanted to engage in scientific research endeavors that contributed to an immediate impact on the lives of others. Being a clinical microbiologist gave me the opportunity to fulfill this desire.

The training program was an intense two-year experience. That primarily focused on learning the technical aspects of the laboratory diagnosis of infectious disease. Infectious disease diagnosis includes many diverse fields within microbiology including bacteriology, mycology, parasitology, virology, and serology. I learned many different skills including the recognition of infectious agents microscopically, by culture, by changes infectious agents caused in tissue culture cells, by detection of microbial virulence factors, and by the immune response that was made against them. I also spent many hours with physicians and would often accompany them during their patient visits. I quickly saw how the work we did in the laboratory impacted on the decisions physicians made in caring for their patients. During the training program I also engaged in translational research, which involves taking ideas developed in the basic science laboratory and applying them to clinical care. Completion of the training program made me eligible to take the American Board of Medical Microbiology examination, an examination I passed a year after completing my training. The American Board of Medical Microbiology exam is administered by the American College of Microbiology, which is a part of the American Society for Microbiology. Those who pass the two-part examination, which includes a three-hour written exam and two-hour oral examination, become Diplomates of the American Board of Medical Microbiology, D(ABMM). The D(ABMM) is a credential that communicates to employers that the individual is competent at the highest professional level in the field of clinical microbiology. At the conclusion of the training program I was fortunate to be chosen as the chief of microbiology at St. Christopher's Hospital for Children. My job at St. Christopher's Hospital was an incredible learning experience and also a humbling experience for me. I was put in charge of a lab with a staff of six and to be honest I really didn't know much about the special microbiology challenges that working in a children's hospital presented. Fortunately for me, the laboratory supervisor, Karin McGowan, was very generous in teaching me the rudimentary knowledge I needed before she went off to graduate school to successfully pursue the same career path I had. Weekly patient rounds with physicians quickly showed me where my major deficiencies in knowledge lay. I spent hours in the library reading the pediatric microbiology literature. But the answers to some of my questions could not be answered there. In particular, we were seeing a microorganism *Pseudomonas* (now *Burkholderia*) *cepacia* frequently in blood cultures and subsequently in autopsy specimens in a unique patient population, children and adolescents with cystic fibrosis. Nothing was written in the literature about this organism as it related to cystic fibrosis. Thus began a more than twenty-year odyssey of studying this organism that continues today.

Currently, I am the director of the clinical microbiology-immunology laboratory and phlebotomy services at the University of North Carolina I am also the director of the training program in medical and public health microbiology at the University of North Carolina. My eighteen years at UNC Hospitals have been wonderful. I am fortunate to work with a highly dedicated and skilled laboratory staff who have been essential to my professional success. I have wonderful colleagues on the medical staff, many of who are internationally recognized experts in infectious diseases and cystic fibrosis, the two disciplines where my research interests intersect. I am involved on a daily basis with current major infectious disease problems. I have been fortunate to be recognized for my contributions to the field having been elected as a Fellow of the American Academy for Microbiology, F(AAM) and having received the Sonnenwirth Award for Leadership in Clinical Microbiology from the Academy. I have also received teaching awards and other honorific awards as well.

One of the important aspects of my job is the creation and dissemination of new knowledge. I have done this in four ways: by teaching, writing, editing, and interviewing with the press. Over the twenty some years of my career I have given over 150 invited lectures throughout North America on various topics as they relate to clinical microbiology. The audience is typically bench level clinical microbiologists, laboratory supervisors or directors, and infectious disease physicians although I have given lectures to the general public as well. I have hosted scientists from around the world who learn the fundamentals of laboratory diagnosis of sexually transmitted diseases, information that they then use in hopes of helping to control the AIDS epidemic in their countries. My colleagues and I have written over one hundred scientific articles, book chapters, and books, many of them on the microbiology issues directly related to the lung disease of patients with cystic fibrosis. I have recently finished the 3rd edition of a textbook entitled *Cases in Clinical Microbiology and Infectious Diseases.* The genesis of this textbook is interesting. My wife and I were bemoaning the fact at lunch one day that fourth year medical students had forgotten much of what they were taught in the first year of medical school about microbiology. We decided not to "blame the victim," the medical students, but rather blame ourselves, the teachers, for not making it clear why the information they were being taught was relevant to patient care. This book attempts to do just that. I currently serve on the editorial board of four of the five major journals in the field of clinical microbiology. In that role, I review papers prior to publication to ensure that the information found there is scientifically relevant and rigorously derived.

II. Description of Job Duties

As the director of the clinical microbiology-immunology laboratories (CMIL) and phlebotomy services in an academic medical center, I wear two hats. One is as the director of a multimillion dollar enterprise that employs approximately eighty people and operates 24 hours a day, 365 days a year.

The other is as an academician in a medical school with the responsibility of teaching, research, and service. It may seem that these activities would be at odds with one another but as we will see, they are complementary activities. As a lab director, I am responsible for the scientific and technical aspects of the diagnostic tests done there to ensure their accuracy and clinical relevance. Since health care is a highly regulated industry, I am also responsible that the work done in the laboratory is in compliance with a variety of federal agencies. The goals of these agencies are to safeguard the care that patients receive, ensure the safety of the individuals who work in the laboratory, and protect the interest of the federal government. If rules are broken the laboratory could lose its license, be fined, or in extreme cases the director of the laboratory could be fined or jailed.

The laboratory director must also evaluate the many "practice guidelines" that are promulgated by federal agencies such as the Centers for Disease Control (CDC) and by voluntary agencies such as the Infectious Disease Society of America (IDSA), the American Society for Microbiology (ASM), or the American Academy of Pediatrics (AAP) to determine if our laboratory is following their guidelines and recommendations. As a laboratory director, I am also responsible for a staff of forty in the CMIL and forty in phlebotomy services. The lab is divided into areas that include diagnostic services for the detection of infections due to bacteria, fungi, mycobacteria, parasites, and viruses. The organization of this endeavor involves individuals with different job responsibilities and skill sets that I will briefly outline.

The backbone of the laboratory is the bench level technologists. These individuals typically have a bachelor's degree in medical technology or microbiology and have passed a speciality examination. In many states, a license is required in order to work in a laboratory. These individuals perform more than 200,000 tests annually in our laboratory. They are in daily contact with physicians often interpreting the test results for them. They must be both highly computer and scientifically literate. They are expected to be competent in several of the diagnostic areas within our laboratory. Their annual salary ranges from $35,000 to $60,000.

Because the CMIL is a complex organization there are three managers. These individuals are all medical technologists with at least ten years of experience. Many lab supervisors have advanced graduate degrees typically in microbiology, public health, or business administration. They too must be licensed. They are chosen for this position based on high levels of technical ability, communication, and leadership skills. They manage the lab on a day-to-day basis. Although I have a consultative role, they are responsible for budgeting, scheduling, evaluating personnel, and maintaining our computer systems. We work together on training of personnel and development of new test procedures. Together, we work very hard to make sure that we have a safe and intellectually stimulating working environment for our employees, that we meet the diagnostic needs of our patients, and that we are in compliance with the many rules that govern how we do this diagnostic work and how the hospital is paid for it. Their annual salary range is $45,000 to $70,000.

How does the role of overseeing a laboratory that generates millions of dollars in revenue mesh with my role of being a medical school professor? Professors are expected to contribute in three areas to the medical school: research, teaching, and service. My role as a director of a clinical laboratory allows me to do all three.

Research

I do clinical research. The research I do is determined by the clinical problems that I encounter in the laboratory on a daily basis. I have already described my long-standing interest in *Burkholderia cepacia,* but I also have worked on clinical research problems with several other organisms. There are several underlying themes in the research I do. First, it is highly collaborative. I have worked with over one hundred collaborators during my career, some of whom I have never personally met. The individuals with whom I have collaborated include a high school student, undergraduate, graduate and medical students, medical technologists, physicians, and other scientists. It is important to recognize that science when done well is a highly collaborative endeavor.

Second, we have a common purpose, trying to solve problems related to patient care. These problems may include better ways to diagnose a particular type of infection, understand how that infection is spread in a population, how best to treat that infection, or as was seen with *B. cepacia,* recognizing the types of infection caused by microorganisms not previously known to cause human infections. Third, most research projects are relatively brief in duration lasting anywhere from three months to two years. Although I have maintained a research interest for more than twenty years for *B. cepacia,* the projects I have done have been self-contained and finite. We have asked a specific question to which we sought a specific answer. In basic research, scientists often follow a problem wherever it might lead and the pursuit of their specific problem may stretch for decades.

Teaching

The most critical part of my job is teaching and I am very fortunate to teach many different learners in many different settings. The most important students I have are the laboratory staff. It is my responsibility that their knowledge is at the cutting edge of clinical microbiology so we can best serve our patients and the physicians that care for them. I teach physicians primarily through phone or face-to-face consultation about specific issues that impact the care of their patients. I teach postdoctoral fellows and residents about clinical microbiology so one day they may, too, direct laboratories. I teach first year medical students and pathology graduate students in the more traditional classroom setting and I have taught undergraduates at UNC from time to time. In addition I am actively involved in continuing medical education as I have already described.

Service

The service aspect of my job is intuitive. I serve the patients cared for in our hospital by ensuring they receive state-of-the-art clinical microbiology and immunology diagnostics services. As the clinical microbiologist at North Carolina's flagship hospital, I also must play a role in protecting the public health of its citizens. This is a role I take very seriously and why I have spent countless hours learning and thinking about the problem of bioterrorism. Additionally, I serve on the medical school admissions committee at UNC for eleven years. Although the irony did not escape me, it was truly a privilege for me to work with others to identify the young people who would become the future physicians in our state.

III. Advantages of Career

Every day at work, I am met with new challenges that test my intellectual and problem-solving skills. I get to work with highly motivated people. I get to have a direct impact on the care of individual patients and also on the public health of the people of my state. I have had the opportunity to travel throughout the world professionally and also to host individuals from many foreign lands with the express purpose of making the lives of our fellow humans better. I learn something new every day that can be directly beneficial to others.

It is important to understand that individuals with my background and training have a number of career options available to them. I consider myself an academic clinical microbiologist. Some of the other opportunities available are to direct a state, county, or city public health laboratory, work for any number of federal agencies such as the CDC, FDA, and NIH, work as laboratory directors in the military, work as directors of large private laboratories, direct a multi-hospital system's laboratory, or work in a number of capacities in the medical diagnostics or pharmaceutical industries. The salary for a doctoral level clinical microbiologist ranges from $60,000 to $200,000+ per year.

IV. Essential Related Skills for Success

With the rapid changes that are occurring in the medical sciences, anyone contemplating a career in this field must be prepared to be a lifelong learner. The microbiology I learned twenty-five years ago is as obsolete today as the typewriter is in the computer age. Much of the information that you will learn during your career must be self-taught. This requires a great deal of self-discipline. It also requires an ability to manage large amounts of information. Computers are tremendous assets in this regard because they allow you to find information very rapidly. This is very important because with scientific knowledge expanding exponentially, it is impossible for most individuals to remember all the information they need. Rather what is needed are good problem solving and data management skills.

As a director of a clinical microbiology lab, there are many management skills that are needed which were not part of my formal education including my postdoctoral training. Through our human resources department, I have been able to take day and weeklong courses that have given me rudimentary management skills. Having a course in business management or industrial psychology would have been very helpful in my career.

Clinical microbiology is rapidly changing due to the molecular biology revolution. It will be necessary in the future that individuals contemplating a career in this field have a working knowledge of the field of genomics and proteomics.

V. Additional Advice

I know that many of you reading might think that I sure have been lucky in my career and you would be right but only partially so. I faced many disappointments along the way and had many tough times. What I will give myself credit for is not letting disappointment and hardship defeat me. I have learned over the years that you can't change the past so when I fail, I recognize that failure and try to determine how I can avoid making the same mistake in the future. I don't fret and worry about past mistakes; they are done, I learn from them, try not to repeat them, and move on.

Finally, I think it is really important to have both mentors and heroes. If you are really lucky, you will have a mentor who is also your hero. You may be asking yourself two questions. First, what is the difference between a mentor and a hero? Second, who needs heroes? Let me answer the second question first. We all need heroes because heroes give us a model for which we can strive. My personal hero is Louis Pasteur. To me, he is one of the greatest scientists who ever lived and did much to improve the lot of humanity. He was an imperfect man, arrogant, egotistical, and neglectful of his family. Is he who I want to be? No, but do I want to try to improve the lot of humanity as he has, absolutely. Will I be able to do this on the scale that he did? Certainly not but what I can do is make sure that what I do professionally is guided by the tenet that I want to do something for the greater good. A mentor, on the other hand, is someone who can help guide you in your career. They should also be someone who you can trust has your best interests in mind. A mentor should serve as a role model but the more important role is to offer constructive criticism as Don Robertson and Larry McCarthy did for me. Having trained eighteen post-doctoral fellows and having countless medical student advisees, I view myself as a mentor for many. Giving constructive criticism is often a difficult task but is the most important thing a mentor can do. It is difficult, not because I find it hard to give it, but the recipient often does not accept it in the spirit it is given. As a result, the recipient is hurt, angry or both because they think their mentor is finding fault with them. Yes the mentor may be finding fault but they are faults that the mentor judges can be corrected and in correcting them, the individual will enjoy even greater success. As a mentor, it is always easier to say nothing rather than be critical. However, a mentor's primary responsibility is to offer constructive criticism to help enhance educational and professional development.

VI. Pitfalls to Avoid

Anyone considering a career in clinical microbiology had better develop a thick skin. One reason is there are essentially no programs in the United States that offer a Ph.D. in clinical microbiology. In my experience, the vast majority of Ph.D. advisors are not enamored of their students going into this field after completing their doctoral degree. Many advisors don't really understand what a clinical microbiologist does and thinks this career path is inferior to the one they have chosen. Another place that thick skin is needed is in dealing with the small percentage of physicians who are rude and confrontational when dealing with the laboratory. What has to be remembered is that the goal is to care for the patient, not prove who knows more about microbiology. Diplomacy is the key when working with physicians.

One of the biggest challenges facing anyone in a career in the medical sciences is the requirement to be a lifelong learner. It is far too easy to become self-satisfied and intellectually lazy especially if one has success early in his or her career. To maintain success in clinical microbiology and in the medical field in general, one must work continuously to keep abreast of the rapid changes that occur in his or her field of endeavor. I think that one thing that I wished I had learned earlier in my career is how to say one of the shortest words in the English language, N-O. Never take on more tasks than you can reasonably handle. If you are skilled at what you do, you will soon learn that your skills are in great demand. Being able to say no allows you to do fewer tasks very well rather than many tasks poorly. This makes your job more satisfying and results in lower stress levels.

Finally in order to do your job well, you have to take care of yourself. Being a clinical microbiologist in an academic medical center is a very stressful job. You need an outlet for stress and you need it on an ongoing basis. Even at fifty, I play basketball three or four times a week as well as hike and swim on a regular basis.

VII. Qualifications Required

Most doctoral level clinical microbiologists have a Ph.D. or Dr.P.H. (Doctor of Public Health degree) but other doctoral level degrees qualify individuals for postdoctoral training in clinical microbiology. Individuals can then do postdoctoral training in clinical microbiology. The committee on Postdoctoral Education Programs through the American Society for Microbiology accredits nine institutions that have training programs in clinical microbiology. Individuals who complete this program are eligible to take the American Board of Medical Microbiology examination. Some individuals are able to obtain positions in clinical microbiology without formal postdoctoral training but it is difficult. A bachelor's degree and master's degree in microbiology, biology, immunology, genetics, and medical technology are minimally required for most entry-level jobs in a clinical microbiology laboratory.

❶ Microbiology is the branch of biology dealing with the study of microscopic organisms and their interaction with other living organisms and the environment.

❷ More information about a career in the clinical sciences can be obtained from: *http://www.naacls.org*.

❸ General information on microbiology can be obtained from: *http://www.microbes.info*.

Source: U.S. Department of Labor. *Occupational Outlook Handbook*, 2002–03 ed. Indianapolis, IN: JIST Publishing, 2002.

Take Home Points

- Clinical microbiologists directly participate in patient care by providing diagnostic services
- Clinical microbiology is an intellectually challenging and personally rewarding career
- The field of clinical microbiology requires good problem-solving and data management skills
- It is very important to have heroes and mentors that positively impact your lives
- Learn how to say "no" and learn how to concentrate your time on activities that will increase your success in all your endeavors
- Ph.D. or equivalent with postdoctoral training and ABMM certification via examination is usually required in order to become a clinical laboratory director. Bachelor's and master's degrees are usually required for entry-level positions

Suggested Reading List

Alibek, K., & Handelman, S. *Biohazard*. New York: Random House, 1999.

Gilligan P., Smiley, M., & Shapiro, D. *Cases in Medical Microbiology and Infectious Diseases*. Washington, DC: ASM Books, 2003.

Peters, C., & Olshaker, M. *Virus Hunter: Thirty Years of Battling Hot Viruses Around the World*. New York: Anchor Books, 1998.

MOLECULAR BIOLOGY

> The difficulty lies, not in the new ideas, but in escaping the old ones, which ramify, for those brought up as most of us have been, into every corner of our minds.
>
> —*John Maynard Keynes*

Dr. Karen Joy Shaw

Team Leader, Infectious Diseases
Research Fellow
Johnson & Johnson Pharmaceutical Research & Development, L.L.C.

I. Biographical Sketch

My first recollection of a profound interest in science was in elementary school, when I picked up my older sister's biology textbook and began to read the chapter on Mendelian genetics. I was fascinated with the concept that one could mathematically predict the color outcome of the mating of cattle based upon the starting genetic make-up of the parents. Later on, in a high school advanced biology class, I remember spending a considerable amount of time proving that the fruit flies we had used in a genetics lab could not have been as advertised, and that in fact we had been sent the wrong genotype for our mating pairs. That was my first experience with refusing to accept dogma when the data generated from an experiment tell you otherwise. I majored in biology as an undergraduate and had decided early on to become a high school teacher.

In my junior year I began my search for a laboratory to work in, my search leading me to the laboratory of Dr. Marion Himes. Here I performed genetic experiments on the marine dinoflagellate *Crypthecodinium cohnii*, a beautiful organism that sparkled like tiny diamonds in the clear liquid culture medium. I spent two wonderful years in her laboratory, over which time she convinced me to stay in research and go to graduate school, and give up the idea of becoming a high school teacher. After my undergraduate research experience, I knew that I wanted to begin my graduate research project right away. I was extremely fortunate that my graduate advisor, Dr. Claire Berg, was sympathetic to my wishes. I began working in her laboratory about three months after starting graduate school. My first experiments in her laboratory involved transposon mutagenesis and characterization of mutants of *Escherichia coli* and *Salmonella typhimurium*. I was once again fascinated by the link between the way an organism looks (its phenotype) and the predictions one could make about the corresponding genotype. Through this process, we were able to identify novel regulatory elements and to define the organization of genes within operons. My research then shifted to the study of a particular amino acid biosynthetic pathway where I learned the power of the complementation between genetics and biochemistry. I published seven papers in my four years in Claire's laboratory. As my graduate career was coming to a close, I began to think about what would be an appropriate postdoctoral learning experience. I felt that it was important to broaden my knowledge and work with a different experimental system. Yeast was the up and coming model for studying higher organisms, and the tools of molecular biology were just becoming established. As a result, I accepted a position in Dr. Maynard Olson's laboratory at Washington University in St. Louis. My research involved a molecular approach to the study of the regulation of the expression of a yeast tRNA gene. I completed two years of postdoctoral training, and then worked for a short time at a small biotechnology company. Seeking a more stable environment, I moved on to a position at the pharmaceutical company, Schering-Plough, where I spent fifteen years gaining knowledge and experience that allowed me to move up within the company. I learned the importance of teamwork, and enjoyed the close camaraderie that was established over this period of time. Our laboratory was one of the leading laboratories studying the epidemiology of bacterial drug resistance. Epidemiology is the branch of science dealing with the mode of transmission, distribution, and prevalence of disease.

For the past three years I have been at my current position at Johnson & Johnson Pharmaceutical Research and Development. Although we are part of a large pharmaceutical company, the work environment has many positive aspects of a small, close-knit company. Due to a relatively flat management structure, people are empowered to contribute to the growth and the strength of the organization. The major goal of my research is to discover novel agents that are active against a variety of critical bacterial infections. This has involved the development and application of new molecular biology techniques such as bacterial microarrays, and the use of genomic approaches to target identification. I have recently edited a book entitled *Pathogen Genomics: Impact on Human Health,* and have published many scientific articles in journals such as the Journal of Bacteriology, Antimicrobial Agents and Chemotherapy, and Molecular Microbiology.

II. Description of Job Duties

Team leaders in the pharmaceutical industry are responsible for directing a group of 10–25 scientists (Ph.D.s, M.S. and B.S.) with the overall goal of finding new therapeutic agents that will have a positive impact on human health.

Research

In the world of drug discovery, scientific research involves both original basic research as well as more directed research to find new solutions to particular problems. Molecular biologists currently have a wealth of information at their disposal, such as the genome sequences of man, protozoa, plants, bacteria, and viruses. We use many new technologies to understand the function of genes and how disruption, change, or augmentation of a function can lead to a new therapy. The tools for studying genes and their functions include cloning (genetic engineering), DNA and protein sequencing, polymerase chain reactions, DNA microarrays (DNA chips), enzyme and binding assays, genetics, cell culture, protein isolation, purification, and microscopy.

A central core of drug-discovery research involves defining a specific molecular target and then developing a test system or assay to find chemicals that will inhibit or enhance the activity of the particular target. My current work involves finding new agents to treat a variety of bacterial infections. We are studying several bacterial cellular components that are fundamentally different between bacteria and man so that compounds we advance to the clinic will only inhibit the bacterial target. I work closely with several medicinal chemists to improve the qualities of the initial molecules that we find, in order to develop compounds that have a high degree of efficacy (in our case, inhibit the growth or kill bacteria), have little or no side effects, and can be given to patients by a desirable route (such as orally or by intravenous injection). We also work with molecular modelers who allow us to view a three-dimensional picture of how a chemical compound may interact with a particular protein or target, and how changes to the chemical may improve this interaction, resulting in a better drug. Collaboration is a fundamental aspect of drug discovery. It is absolutely critical to continually seek out others who may have information, insights, or skills that will help to rapidly advance a project. As an example, we also consult with scientists in other specialties, such as pharmacology and clinical research, to obtain additional input.

Communication

Communication is a fundamental part of all scientific endeavors. This can involve participation in laboratory meetings, where experiments are discussed and initial findings are reviewed. Other internal meetings may review the progress in different departments, or explore potential collaborations between groups. Often, formal presentations are made to describe the goals and accomplishments of a group.

In the pharmaceutical industry, once a new discovery is made, it is often important that patent applications are filed so that the company can establish ownership of the discovery. This involves working with a patent attorney to define the scope of the invention (perhaps a particular chemical compound or set of related compounds) and what it can be used for.

Scientific findings are often communicated to the scientific community in the form of research articles that are published in scholarly journals. In addition to writing manuscripts for publication, I am often called upon to review the quality of manuscripts that have been submitted by other investigators and offer suggestions and comments to the editor of the journal. Scientific findings are also frequently presented as a lecture or a poster at a scientific meeting. In my field, the American Society for Microbiology holds two meetings per year in which scientists from around the world present their latest research data.

Mentorship

Team leaders provide guidance and mentoring to their staff of scientists. This involves setting the goals for the department, discussing strategies to achieve those goals, reviewing experimental procedures, trouble-shooting experiments, and providing advice and training. I have had the opportunity to train postdoctoral students, graduate students, and other scientists.

III. Advantages of Career

The advantages of becoming a scientist working for a pharmaceutical company are as varied as the people themselves. Most of us are attracted by the idea that our scientific research is oriented towards a goal that provides for the betterment of mankind, whether it be cures for diseases or alleviation of pain and suffering. In addition, it provides an environment where working together as a team is essential, and interacting with others across disciplines leads to the enhancement of our own and the company's goals.

The biggest advantage is that one never stops learning. There are always new discoveries, new technologies, new challenges, and an environment where one can apply the knowledge gained to solving important problems in human health. One delightful aspect is that it is easy to find someone down the hall, across the country, or around the world that you can engage in your quest for scientific discovery.

Another career advantage is having many opportunities to give back to the community as well. I have taken great pleasure in providing scientific demonstrations in elementary schools, participating in high school career nights, mentoring elementary and junior high school students through the wonderful "Science-by-Mail" program out of Liberty Science Center in New Jersey, working with students in invention competitions, and guest lecturing at colleges.

IV. Essential Related Skills for Success

The fundamental skills as a molecular biologist in industry are the comprehensive knowledge in a particular scientific discipline such as microbiology, and a passion to continue to learn. Other key skill sets include computer proficiency, critical thinking and problem solving, and knowledge of statistics. Many computer courses are available, and I have found that having additional skills in basic computer programming and logic have been valuable. A high emphasis is placed on communication, and thus skills in both public speaking and technical writing are critical. Participation in debate teams or toastmasters clubs can provide a forum for learning to be comfortable speaking in front of large audiences. Due to the fast pace of the industry, time management and multitasking skills are important as well.

In terms of career advancement in industry, leadership and interpersonal skills are essential. These include getting along with people with diverse backgrounds and interests, the ability to negotiate to create a "win-win" solution, providing and accepting candid feedback, and learning how to listen. Joining teams or clubs and holding leadership positions can provide a forum for developing these skills.

V. Additional Advice

The only way to determine if a scientific career is right for you is to try it out as early and as often as possible. Both high school students and undergraduates can find opportunities to volunteer or obtain internships at local colleges or universities, or within the pharmaceutical industry. In my laboratory, I mentored two high school juniors on their research projects when they participated in the Westinghouse Talent Competition (now called the Intel Science Talent Search). As an undergraduate, I volunteered in a laboratory for two years and was able to obtain a scientific publication of my work. Many colleges and universities have independent study programs where college credit can be obtained for this work. The key to these studies is not necessarily the specific research project that you work on, but learning about the scientific process. The goal is to understand how research is performed, to learn to critically evaluate your experiments, to understand the ups and downs of research, and to determine if this is something that you find personally rewarding. In addition, these experiences provide mentors and contacts that will be important references for your career. Pharmaceutical companies often provide summer internships to college juniors and seniors. These are wonderful research opportunities that will also provide a glimpse at what life in industry is like.

VI. Pitfalls to Avoid

If you ask someone off the street about a scientific profession, the general opinion is that it is a lonely pursuit. Nothing could be further from the truth. Too many brilliant scientists have failed in industry because they have tried to work in a vacuum or under a veil of secrecy. Those that succeed generally do so because they have created a network of people that they can consult with, and will readily seek collaborations to advance their work.

Often times students will become focused on their scientific goals and decide that other subjects are less relevant. In the span of a career, it is impossible to predict what will have been important or unimportant in your studies. Therefore, it is critical to obtain a well-balanced education. Two of the most valuable courses that I took in college were a *Philosophy and Logic* class and an English class entitled *Greek and Latin Roots*. Science has changed significantly in the past twenty-five years, but putting together a logical argument and understanding the origins of words have stood the test of time.

VII. Qualifications Required

A Ph.D. is generally required for a team leader position in industry. Postdoctoral experience and an additional 7–10 years of industrial drug-discovery experience are also required. A bachelor's degree in a science-related field is generally required for an entry-level position in the pharmaceutical industry.

ⓘnteresting Facts

❶ More information on a career in the pharmaceutical sciences can be obtained from: *http://www.aaps.org*.

❷ Molecular biology is the branch of science that deals with the chemical properties and organization of genetic material that give rise to proteins in living systems.

❸ There are currently great career opportunities for molecular biologists.

Source: U.S. Department of Labor. *Occupational Outlook Handbook*, 2002–03 ed. Indianapolis, IN: JIST Publishing, 2002.

Take Home Points

- Team leaders are responsible for directing the overall research of a group of scientists towards the goal of discovering and developing new drugs of therapeutic value
- Pharmaceutical research is extremely rewarding work that is oriented towards improving human health by providing cures for diseases or alleviating pain and suffering
- Critical thinking, problem solving, computer proficiency, public speaking, technical writing, and "people skills" are all essential skills for success as a team leader
- Try out your scientific career early by taking advantage of volunteer and internship opportunities
- Obtain a well-balanced education—you'll never know what you'll need in the future
- A Ph.D. and additional drug-discovery experience are necessary qualifications for a team leader position in industry. Typically, a B.S. in a science-related field is an entry-level educational requirement for the pharmaceutical industry

Suggested Reading List

Hugo, W., & Russell, A. *Pharmaceutical Microbiology*. Oxford, UK: Blackwell Publishers, 1998.

Stonier, P. *Discovering New Medicines: Careers in Pharmaceutical Research and Development*. Hoboken, NJ: John Wiley & Sons, Inc., 1994.

Werth, B. *Billion-Dollar Molecule: One Company's Quest for the Perfect Drug*. New York: Simon & Schuster, 1995.

NURSING

> I will do all in my power to maintain and elevate the standard of my profession.
>
> —*Florence Nightingale*

Mrs. Valerie J. Shereck

Adult Nurse Practitioner
Penrose-St. Francis Senior Health Center
Centura Health

I. Biographical Sketch

My first career goal did not involve a science-related field. My first career goal was actually to become an English teacher. This all changed when as a junior in high school I watched a film depicting the birth of a child. Viewing this film was such a moving experience for me that I felt I wanted to pursue a career where I could actively be involved in the process of promoting life. I was not really prepared as a young woman of eighteen for the complexities involved in this career choice. I had never volunteered or worked in a hospital setting. In fact, I once almost fainted when my sister was receiving multiple injections prior to our moving to Germany. There was so much to learn such as drug calculations, chemistry, biology, microbiology, anatomy, and physiology. Also included in the curriculum were the social sciences, such as psychology, sociology, nursing history, and nursing theory. Nursing research was not stressed at this time but was so in the latter part of my education including my BSN, MSN and post-master's program.

I initially went to a hospital diploma program, Seton School of Nursing at Penrose Hospital in Colorado Springs, Colorado, which was associated with the University of Colorado at Colorado Springs. It was a three-year intensive course that involved a great deal of "hands-on" training. As I look back on my career this training and educational experience was invaluable to me. These diploma programs are obsolete today and are now replaced with the four-year baccalaureate programs as well as two-year community college programs.

During this period of my education I learned a great deal about the human body and the human mind. I found the study of the human body to be extremely intriguing as the body is a wonderful, intricate machine which seeks homeostasis or balance, both physiologically and psychologically. I still marvel each day at what I learn about each individual patient and his or her response to their environment.

I graduated from nursing school in 1968 and worked in medical surgical nursing as well as intensive care, coronary care, cardiology and oncology. I also worked in management as the director of nursing at an alzheimer's center. I pursued additional education in the form of a bachelor's in nursing which I received in 1990 from Beth-El College of Nursing. I then completed my masters of science at Beth-El College of Nursing. Shortly, after I received my post-master's certificate as an adult-geriatric nurse practitioner from University of Colorado at Colorado Springs. Currently, I am certified by the American Nurses Association as an adult nurse practitioner.

After graduating with my master's degree, I was able to expand my nursing horizon and have greater autonomy. I began working part-time as a community case manager and also part-time as a clinical instructor in the area of medical surgical nursing. Both of these positions were exciting, challenging, and rewarding and offered me the autonomy and independence that I was seeking. One of my favorite areas of nursing is that of patient education. I enjoy teaching patients preoperatively about what to expect before and after their surgical procedures. As an oncology nurse, patient education was an essential part of my job. In my current position as a nurse practitioner I have developed an extensive assortment of teaching aides for patients dealing with a variety of health-related topics.

As a result of these early educational experiences I formed the hospital's current Patient Education Committee. This committee has grown to be a very effective, multidisciplinary educational team. I was also facilitator for the American Cancer Society's "I Can Cope" program, which is an eight-week educational program designed for patients and their families. I received an award for my role as facilitator from the American Cancer Society in 1993.

I have written and published numerous articles for both the hospital bi-monthly newsletter and the hospital's quarterly nursing newsletter. These articles have been on a variety of topics to include alternative medicine, the role of the Advanced Practice Nurse, and medication compliance in the elderly.

I am active in numerous nursing organizations. For example, I am currently am a member of the local and national Oncology Nursing Society. I am also a member of the Southern Colorado Advanced Practice Nurses Association and a member of Sigma Theta Tau International XI PHI Chapter. Moreover, I am immediate past president of the Colorado Nurses Association.

My current position is that of adult geriatric nurse practitioner at the Penrose-St. Francis Senior Health Center. This program is designed to help provide care for medicare patients who are in need of primary care and are unable to find a physician. In our geographical area, physician reimbursement for Medicare patients is extremely low, and at the time we started our program only two physicians in the community were accepting new medicare patients. Our model is a unique one in that the nurse practitioners are the primary care providers and the physicians, who only work part-time act as consultants.

II. Description of Job Duties

Primary Patient Care

My principal role as a primary care provider is to take an adequate history of the patient's chief complaint, to assess the patient physically, make a clinical correlation between their reported symptoms and history with the physical exam, and then make a differential diagnosis. Sometimes this is not a clear cut matter as the patient often has vague symptoms which can include many differential diagnoses and often patients do not include all the pertinent information needed. There are also always numerous psychosocial issues which may cloud or obscure the objectivity of these chief complaints. When the diagnosis is not clear cut, further testing to include laboratory studies, x-rays, and other tests, is often needed to establish the exact cause of the patient's concern. Included in this role is prevention as well as health maintenance. Routine screening tests such as mammograms, flu vaccines, and colonoscopies are recommended at regular intervals to ensure that potential risk factors are not neglected. Patients with multiple medical problems such as diabetes, heart disease, and emphysema are followed up on a regular basis to ensure compliance with their medications, diet, and exercise. Each patient's plan of care must always be individualized and tailored to his or her needs.

Research and Continuing Education

It is essential as a primary care provider for the elderly to be constantly updated on the latest research in not only health maintenance but also in the chronic disease processes such as hypertension, diabetes, osteoporosis, emphysema, chronic bronchitis, and arthritis. It is imperative to read journals as well as other publications describing the latest research findings. The Internet is another source of good information if used wisely. The patient of today, especially the elderly, has access to a multitude of information regarding health care and alternative medicine. As providers, we have an obligation to our patients to be well informed about the latest research and technology available to them and help them make the best choices for their health care.

Education and Counseling

As stated previously in my biographical sketch patient education is very important to me. When patients come to the office they are often anxious regarding the outcome of their visit. They are often frightened about hearing "bad news" about their health care. It is known that during the educational process only twenty percent of what is taught is actually learned. Therefore, a handwritten note with instructions, a typed handout describing the disease process or medication, or a description using a model or picture are reinforcements to the actual process of patient education. I also often give my patients copies of their diagnostic tests with additional handwritten notes. It is also important to counsel patients about preventative medicine and how they can maintain an optimal quality of life through diet and exercise. If the health care provider maintains a good weight, exercises regularly, the patient is often much more likely to follow his or her advice regarding smoking cessation, weight management, and exercise.

Collaboration and Referrals

As a primary care provider for an elderly population who often have complex, multisystem diseases, I often rely on the expertise and skills of my colleagues from my fellow nurse practitioners to knowledgeable physician consultants as well as other physician specialists in the community. I also try to have a collaborative and respectful relationship with other disciplines, including physical therapy, home health care, hospice, pharmacists, and social services. I ask for feedback from my patients after they have been to a specialist or have had physical therapy to ensure that my referrals reflect the quality of care that my patients deserve.

Lastly, it should be pointed out that the major difference between a nurse practitioner and a registered nurse is that nurse practitioners have to meet more rigorous educational and clinical requirements and therefore have additional responsibilities and competencies.

III. Advantages of Career

The role of a nurse practitioner has limitless possibilities. Nurse practitioners are in very diverse settings such as psychiatry, allergies and asthma, women's health, cardiology, gastroenterology, and oncology. Medicine is never dull as there are new treatments, new medications, new research, and new technology being developed everyday. The work is exciting, challenging, and rewarding as each individual patient is unique. I feel that the most rewarding aspect of my job is when a patient returns and informs me my treatment plan has indeed improved the quality of his or her life. There can be no better reward.

IV. Essential Related Skills for Success

In addition to a strong background in the basic sciences such as chemistry, microbiology, anatomy and physiology, and biology a nurse practitioner should also have a firm foundation in the social sciences to include psychology and sociology. Additional expertise or knowledge is needed in pharmacology, nursing research, statistics, nursing process, and nursing theory. I believe that all nurse practitioners need analytical skills in order to make a differential diagnosis as well as critical thinking, creative thinking, and good problem-solving skills. A nurse practitioner should have good organizational skills, and be able to work well with others. He or she should be able to work well under pressure and be capable of multitasking. Basic computer skills are necessary as well as excellent communication skills, both written and verbal. Strong interpersonal skills are very necessary along with the ability to work well with numerous individuals with a variety of job skills, such as medical assistants, receptionists, medical records clerks, and file clerks. I would strongly recommend a course in creative thinking. I took this course while pursuing my master's degree and found it to be invaluable in helping me both with my critical thinking but also in assisting me in becoming more intuitive about how other's think.

V. Additional Advice

Goal Setting

My father, who is a retired colonel in the U.S. Army, taught me the value of setting goals. Even today, at the age of eighty-three, he still has a flip chart posted in his den with his goals written out for each year. My father was the great motivator for me to go on and pursue my additional education, especially that of being a nurse practitioner. I set goals for myself each day. I usually write them down and eliminate them off the list when completed. I also have a five-year, ten-year and fifteen-year goal plan. I refer to these goals periodically. Having these goals written down helps me to realize that these goals are closer than I think.

Participation in Professional Organizations, Clubs, Meetings

I did not become a member of a professional organization until I started my bachelor's of nursing program in 1986. By that time I had been a nurse for eighteen years. I never realized what I was missing. I first joined the Colorado Nurses Association, local chapter. I then became very involved along with a few others in forming a local Oncology Nurses Association. I was a founding member of that organization and was the first treasurer. Today, I am active in and belong to several professional nursing organizations. The value of the networking with colleagues, the value of learning more about other roles in nursing and expanding our knowledge base through this networking is invaluable. Additionally, there are frequently educational programs offered through these professional organizations at reduced or minimal cost.

Mentors

In my career, I have had excellent mentors and also have been a mentor to many individuals. Today, I still have people in my life who act as mentors both professionally and personally. I think it is extremely important to select individuals who are willing and able to help you reach your goals.

VI. Pitfalls to Avoid

As stated previously in my biographical sketch, I noted that prior to entering nursing school at the age of eighteen, I was unprepared for the complexities involved; I do believe that if I had to do it all over I would have volunteered in a hospital, nursing home, or clinic prior to entering nursing school.

I also feel that I did not have a strong background in outpatient medicine prior to starting my career as a nurse practitioner. My primary area of clinical expertise was acute care and this is very different from the standard medical practice I worked in at the family practitioner's office during my clinical rotation as a nurse practitioner.

It is important to remember as a student of nursing to take advantage of any opportunity to perform and enhance your skill base while under the supervision of an instructor. Students are often shy, embarrassed with their lack of knowledge and skill. The message here is to take full advantage of every learning opportunity and forget your inhibitions.

VII. Qualifications Required

Qualifications for nurse practitioners vary from state to state according to licensure agreements. Typically, a bachelor's of nursing, master's of science in nursing, and post-master's nurse practitioner certificate are required. In some states additional education in pharmacology and certification from a nationally accredited organization are also required prior to licensure. In Colorado, after graduation from an accredited NP program, a license will be issued but an additional 1800 hours of supervised clinical experience by a physician is needed prior to obtaining prescriptive authority in the state of Colorado. Prescriptive authority also varies from state to state. Please contact your local state board of nursing for a complete list of the appropriate requirements for licensure to become a nurse practitioner.

❶ Due to the critical shortage of nurses, the nursing profession will provide excellent job opportunities for aspiring nurses.

❷ The nursing profession represents the largest heath care occupation.

❸ More information regarding a career in nursing can be obtained from: *http://nurse-recruiter.com.*

Source: U.S. Department of Labor. *Occupational Outlook Handbook*, 2002–03 ed. Indianapolis, IN: JIST Publishing, 2002.

Take Home Points

- The major difference between a nurse practitioner and a registered nurse is that a nurse practitioner must meet more rigorous educational and clinical requirements to become certified
- Nursing is a dynamic, ever expanding field with a multitude of opportunities. Nurse practitioners have the opportunity to work in a variety of settings
- Success as a nurse practitioner is related to a sound knowledge base in the basic sciences and social sciences
- Critical thinking, organizational ability, creative thinking, analytical skills, problem solving, and basic computer skills are paramount for a successful career in nursing
- It is extremely important to set goals for yourself and to be involved with professional nursing organizations as well as to establish strong ties to mentors
- Qualifications for nurse practitioners vary from state to state according to licensure agreements. Contact your local state board of nursing for a complete list of the appropriate requirements for licensure

Suggested Reading List

Benner, P. *From Novice to Expert: Excellence and Power in Clinical Nursing Practice.* Menlo Park, CA: Addison-Wesley Publishing Co., 1984.

Eagles, Z. *The Nurses' Career Guide : Discovering New Horizons in Health Care.* San Luis Obispo, CA: Sovereignty Press, 1997.

Hawkins, J., & Thibodeau, J. 3rd ed. *The Advanced Practitioner: Current Practice Issues.* New York: Tiresias Press Inc., 1993.

PHARMACY

> Hold fast to dreams, for if dreams die, life is a broken bird that cannot fly.
>
> —*Langston Hughes*

Mrs. Natasha J. Johnson

Registered Pharmacist
Pharmacy Department
Eckerd Drug Company

I. Biographical Sketch

"Help me, help me!" screamed a mother, her child having just fainted. She had just left with prescriptions I had filled for her and her child, but now she was back, clearly distressed. As a pharmacist, there are times when you have unexpected occurrences, but you are thoroughly trained to handle them. "Call the ambulance," I patiently directed the technicians. "Miss, explain to me what happened," I said. She responded, "I don't know, he just all of a sudden fainted, what did I do." So at that time, I had a nervous mother and a little kid lying on the floor. Immediately I checked to see if he needed CPR, but he did not, so at that point, I decided to start a conversation with the kid to keep him alert, and the mother to calm her down. I sat on the floor with his head in my lap, talking to him trying to get him to open his eyes. In the meantime the ambulance arrived and took care of the situation from there.

From the time that I was a little girl, medicine always intrigued me. Whenever I was asked, "What do you want to be when you grow up?" I would reply, "A doctor." All I knew was I wanted to cure the world of all their illnesses. Growing up, I was greatly inspired by my Aunt Connie. She is a registered nurse and has influenced my respect for all health care workers. As a child I was enlightened by her stories, which detailed how she helped people on a daily basis. The older I became, the more I steered away from becoming a physician and was directed towards the fascinating world of pharmacy. The principal reason for this transition was the presence of many malpractice lawsuits. People seemed to have lost their faith and respect for physicians. Another reason for this transition was the inspiration of a profession that would not consume all of my time, and grant me an opportunity to start a family. Physicians work exhaustive hours, decreasing the quality time spent with their families. Family will always be a priority in my life and pharmacy just seemed to be a perfect practice for me to devote my time to. I would still be able to study medicine, help people, begin a family, and make a substantial amount of money all at the same time.

I matriculated in the Philadelphia College of Pharmacy and Science with scholarships and grants due to my academic achievements throughout high school. I was a member of the National Honor Society because I was in the top ten percent of my class, with honors in Latin and Spanish. I started college the day after high school graduation because I had to participate in a scholarship program called PREP (Preparation Retention Educational Program). This is a program for which students were selected on account of their outstanding academic performance and major. Out of one hundred candidates, sixteen were chosen to receive this honor. We were taught the basics of what to expect in college and various professors taught the first semester of school in a six-week time period. After the completion of this demanding summer program I knew the pharmacy curriculum would be intensely demanding but I accepted the challenge, thus determined to finish the curriculum and become a pharmacist.

The curriculum includes various science disciplines and certifications. I was exposed to extensive organic chemistry, physics, and biology courses. Organic chemistry, the study of carbon-containing compounds, elaborates on chemical structures and their functional groups. Within these functional groups, drugs are classified based on their reactivity. Physics courses focused on the laws of thermodynamics and physical forces that affect reactions of drugs. Biology courses emphasized the anatomy, physiology, and neurobiology of the human body and how various chemicals interact and affect the body. In addition to science courses, psychology and sociology courses were also instrumental in my career development. These two disciplines gave me the ability to counsel patients based on their individual and cultural history. Moreover, my training led to the various certifications such as pneumococcal and influenza vaccinations, CPR, and first aid.

In the summer, while an undergraduate, I participated in an internship in an actual pharmacy to get an inside view of the profession. My first pharmacy job was in an independent pharmacy, Home Drugs, where I spent most of my time servicing the customers. Most of the customers became very fond of me and would ask my opinion about certain medications. It was during my summer internships that I discovered my fondness to work in the retail environment. I gained further work experience at Chester County Hospital, Pathmark, and ACME as an undergraduate. Required within the curriculum were three clinical rotations, which included hospitals and retail stores nearby. The goal was to offer students several experiential endeavors guided by a preceptor, a professional that mentors and teaches what is necessary in a particular field of study.

II. Description of Job Duties

In general, the pharmacy profession can be described as the practice of preparing, preserving, and dispensing medical drugs. The average day of a pharmacist in a retail environment can consist of several different tasks, such as reviewing prescriptions from the preceding day, interpreting and filling prescriptions, communicating with insurance companies to solve problems, verifying and dispensing prescriptions, and taking inventory of narcotics.

I start my day in the morning, checking the refills left overnight on my computer. My computer system is designed to keep track of prescription refills ordered throughout the day to be picked up at designated times by our customers. Before I can pull up anything on my computers I have to sign on to all four different screens on each of my three computers. My pharmacy is set up in a manner that allows customer access in three different designated areas. The first window is the "drop off" window where prescriptions are dropped off to be filled. The second window is the "pick-up" window where customers can pick up their prescriptions that are already filled. The last window is the drive-through window. I am expected to answer the telephone from physicians, check prescriptions, and monitor any possible drug interactions all at the same time. It can get very frustrating at times, but you have to learn how to remain patient and in control at all times.

Interpreting and Filling Prescriptions

It is very important to keep in mind that people's lives are in the hands of the pharmacist filling their prescriptions. Interpreting and filling these prescriptions is the most important aspect of this profession. I come in contact with handwriting from physicians that is severely illegible and sometimes resembling Egyptian hieroglyphs. Often times the drug name or strength is written incorrectly and it is my job to call and verify every error written on the prescription. This is the moment that all drug knowledge comes into play. For example, doctors commonly write for a drug combination of Ultram (an analgesic) and Flexeril (a muscle relaxant) which if used together can increase a person's chances of having a seizure. I counseled a customer on this interaction and she told me that she had a strong family history of seizures, so immediately I called the physician to change the medicine. She was very grateful for my intervention and now she consults with me before taking other medication. As a health care professional you must have your patient's best interest in mind every time. If the strength is too high or the medicine interacts with other medications that the patient is currently taking, it is the responsibility of the pharmacist to contact the patient and the physician.

Insurance Troubleshooting

When entering the prescription into the computer the claim must be sent electronically to the insurance company. This can be one of the most difficult tasks of the retail environment. Typically, I experience rejections with expensive prescriptions. Most of the time, rejections occur because insurance companies want the customer to try a less expensive treatment, which may not be in the patient's best interest. In certain situations the physician has to be notified and must submit forms to the insurance processor in order for a prescription to be accepted. This type of situation can be troublesome and time consuming for the pharmacist, the doctor, and the patient. It usually takes days to resolve these problems, thus prolonging the chance to treat the patient. Fortunately, in my pharmacy, I have an insurance specialist who handles a majority of these types of problems.

Verifying and Dispensing Prescriptions

The liability of checking filled prescriptions before they are dispensed to the customer is an important process. I must make sure that it was filled under the correct patient by verifying date of birth, sex, address, telephone number, and other important information. Then I have to make sure that I possess the correct medicine in the labeled bottle. Once the bottle is labeled with my initials on it, the responsibility behind the contents is that of the pharmacist.

Narcotic Prescriptions and Inventory

Some of the other duties of the retail pharmacist require me to handle, dispense, and maintain inventory of narcotics. There are strict laws and restrictions governing these issues. It is very important these laws are practiced because the board of pharmacy performs unexpected audits in pharmacies. During these audits the handling of narcotics is emphasized. Some of the major issues are the storage of narcotics as well as the proper documentation of each filled prescription. If anything is handled incorrectly there can be heavy fines and other penalties.

III. Advantages of Career

There are countless advantages of a career in pharmacy. Once you earn your pharmacy degree, the job arena opens up to a world of fabulous opportunities. For example, you can be a nuclear pharmacist instead of working in retail or hospital settings. Many pharmacists enjoy counseling patients in nursing homes, positions in industry, or marketing. Within each domain of pharmacy, you may become an assistant manager, supervisor, or district manager. One of my favorite areas of my occupation is the interaction with the customers. I enjoy speaking to people about their health and prescriptions. The community relies on my expertise for drug information. It is a great feeling when you save a person from a life threatening reaction by sharing your drug knowledge.

The rewards and benefits that I acquire from my career are endless. Socially, I have an opportunity to meet new and exciting people on a daily basis and I am able to build a rapport with the community. Financially, as a pharmacist you can earn a very competitive salary. In most areas you can make a great living just working part-time. With a high demand for pharmacists, there exists a high supply of wealth. Companies will just about bid for your services, and you can have sign-on bonuses averaging $5,000. Some bonuses are lower, some are higher ($10,000). If you are family oriented, the pharmacy profession gives you the opportunity to spend more quality time with your family. Personally, direct customer contact strengthens my communication skills and my ability to utilize my education in practical situations, builds a strong respectable reputation for health care professionals, and gives me a true sense of purpose. From all of the benefits of being a pharmacist, I am able to live comfortably, enjoy long exciting vacations, and have time to do whatever I desire.

IV. Essential Related Skills for Success

Your education will provide you with a sufficient background for drug knowledge; however, it takes a tremendous amount more to succeed as a pharmacist. Many of the skills needed to operate effectively are not taught in the classroom but learned through experiences. Some essential skills are the application of your knowledge to everyday occurrences, communication, time management, leadership, analytical, observational, problem-solving, organizational, and counseling skills. As a pharmacist, you will be in many situations where you will need to apply your drug knowledge and teach patients how to take their medication. Nowadays people are much more aware of their health and they want to know exactly how the medicine works in their body and when they should experience some effect. You will be expected to speak and understand medical terminology when conferring with a doctor or nurse. When counseling your patients, explain everything conducive to their understanding.

Leadership skills are one of the most important skills when operating a pharmacy. You are the leader of your team when you become a pharmacist. The pharmacy cannot maintain efficiency without teamwork, which includes your clerks, technicians, and management. With your organization and direction, your team will function properly and very productively. The operations performed throughout the day must be communicated well to your workers. The formula for customer satisfaction depends upon leadership. A non-stressful working environment creates efficient workers, which results in excellent customer satisfaction.

V. Additional Advice

In order to make a smooth transition from college to the professional environment, I strongly recommend that you search for an internship before you graduate. Internships play a critical role in your training to become a pharmacist. Internships concentrate on teaching you how to apply your knowledge and function efficiently in a professional environment.

I was chosen to participate in an internship in my junior year of college. I was given a home store where I was able to share in the workflow of that pharmacy. Under the supervision of my preceptor, I counseled patients and exercised my drug knowledge to perform other duties as well.

I also had the opportunity to be involved in community service during this internship. My preceptor and I set up an educational tour for diabetic patients. During this tour, we described the importance of interpreting the percent daily values or nutritional facts listed on certain foods of each major food group. Although not all internships pay students for their time, the experience in itself is worth it. After the completion of my internship I was certain that I had chosen the perfect profession.

Also, join various professional organizations, so that you may gain a better understanding of what's going on in your profession and how you can prepare yourself better. I am an active member of the Virginia Pharmacists Association and American Pharmaceutical Association. I have been a member of the American Pharmaceutical Association since my second year of college. These organizations supply me with important updates on new and old drugs as well as the changes in pharmacy law. They have annual meetings in which many pharmacists from all over the world get together to discuss major issues in pharmacy.

To all students, it would be useful to take a communications or speech class as an elective as early as possible. It will prepare you to become confident and a productive communicator. Superb communication skills will benefit you whether you are giving a presentation in front of an audience or just engaging in one-on-one counseling.

VI. Pitfalls to Avoid

Unfortunately, there are potential obstacles that students may encounter during their educational and professional career. One of the most difficult barriers faced during your education is balancing your course load with working outside of school. Most students find it hard to work while they are taking such demanding courses. Many of my fellow peers came close to failing out of school because they had to work to pay for their room and board. The best way around this problem is to condition yourself to study on a daily basis: time management. Working to help pay for your expenses is difficult to avoid, but your priorities should focus on your education. Personally, if I could change anything about my college career, I would have decreased the amount of time spent working. Prevention of these pitfalls can be accomplished while you are in high school. As a senior, you should concentrate more on conditioning yourself for intense academics. Take challenging courses that will force you to study everyday. Carefully consider your financial obligations for college. Do not choose a college that is very expensive when realistically you may have trouble paying tuition.

On a professional level, once you are a registered pharmacist your attention will be diverted to multitasking. Like I mentioned in *Descriptions of Job Duties* as a pharmacist you will be expected to perform many tasks at one time. Frequently when I am in the process of checking a prescription I am interrupted to answer the phone or counsel a patient. Those are the situations that increase your chances of making a mistake. Prescription errors are a pharmacist's worst nightmare. Unfortunately, we are human and we make mistakes, but please remember to take your time and concentrate to prevent these occurrences. As a pharmacist you are responsible for the prescriptions leaving your pharmacy so make sure you verify all the information correctly. The extra seconds that it will take you to recheck a prescription will save you from a lot of restless nights because of an error that could have been avoided. Another obstacle within the pharmacy profession is stress and becoming frustrated. Develop stress-reliever strategies to combat the inevitable problems associated with the work environment.

VII. Qualifications Required

The bachelor of science degree in pharmacy is required to work in any field of pharmacy. It takes a minimum of five years of college including three professional years of pharmacy curriculum to obtain the degree. Recently, the pharmacy curriculum has changed making it mandatory to graduate with a Pharm.D. degree, which adds an extra year of school.

Once your classes are completed you will be expected to fulfill a series of clerkships that will give you experience in several different domains of pharmacy. Once you graduate, you will have the chance to choose a fellowship or a residency in a specified division in pharmacy. A fellowship usually lasts two years and focuses on students who want to pursue a career in research. A residency usually lasts one year and helps you focus on a specialty area such as diabetes, infectious disease, or asthma.

Those who wish to start working immediately will be expected to complete a specified amount of hours in a pharmacy after graduation according to the state in which you plan to be licensed. To become a registered pharmacist, you are expected to pass an examination in drug information and pharmacy law, executed by each state. As a registered pharmacist, the year after you receive your license you will have to acquire fifteen hours of C.E. (continuing education) to maintain your license. By law you will need to keep your certificates of passing scores in your pharmacy at all times for DEA (Drug Enforcement Agency) audits. Even though your college education may be completed, you will always be expected to keep up with new drug information.

(i)nteresting Facts

❶ Excellent employment opportunities are expected for pharmacists in the next decade.

❷ More information about a career in pharmacy can be obtained from: *http://www.aacp.org*.

❸ Eighty-two colleges of pharmacy are accredited by the American Council on Pharmaceutical Education.

Source: U.S. Department of Labor. *Occupational Outlook Handbook,* 2002–03 ed. Indianapolis, IN: JIST Publishing, 2002.

Take Home Points

- Retail pharmacists are responsible for filling and dispensing prescriptions, insurance troubleshooting, maintaining inventory, and counseling patients
- There are many promising job opportunities in the field of pharmacy
- Communication, time management, leadership, analytical, observational, problem solving, organizational, and counseling skills are imperative to succeed as a pharmacist
- Internships provide you with practical experience in a professional atmosphere
- It is very important that you remain focused on your principal educational and career goals
- A minimum of a bachelor of science degree in pharmacy, and successful completion of the state boards are required to become a registered pharmacist

Suggested Reading List

Gable, F. *Opportunities in Pharmacy Careers.* Lincolnwood, IL: VGM Career Horizons, 1998.

Gosselin, R., & Robbins, J. *Inside Pharmacy: The Anatomy of a Profession.* Boston, MA: Technomic Publications, 1999.

Peterson, S. *Preparing to Enter Pharmacy School.* Manchaca, TX: S. Swift Publication Co., 1978.

PHYSICS

> There's a difference between knowing the name of something and knowing something.
>
> —*Richard Feynman*

Dr. Mark F. Vondracek

Physics Teacher and Research Advisor
Science Department
Evanston Township High School

I. Biographical Sketch

My parents always tell me that the one thing that stood out when I was young was the number of questions I would ask them about anything and everything I observed. It always seemed natural and right to try and figure out the world in which we live, always trying to quench the thirst that curiosity caused. I was and still am consumed with wanting to learn something new each day, and I am now privileged to be able to share my love for science and learning with over one hundred youngsters every year. I am able to do this because I am a science teacher in a public high school.

Throughout my schooling science and mathematics were always fascinating, and I knew from an early age that I wanted to pursue physics, the most fundamental and mathematical scientific discipline. After graduating from Streamwood High School in a Chicago suburb, I went to the University of Illinois, Urbana-Champaign, to study physics. The dream at that time was to ultimately earn a Ph.D. in physics and become a professor at a major research university in order to teach physics and stay on the cutting edge of research. My main interest in physics has always dealt with high energy particle physics, where the goal is to answer one of the ultimate questions in the universe: What is the world made of? I worked on studying particles that are responsible for some of the fundamental forces observed in nature by becoming a member of a large collaboration at Fermi National Accelerator Laboratory. However, as I worked my way through graduate school and my doctoral research at Fermi National Accelerator Laboratory, I realized that although the research was exciting and interesting, my heart was elsewhere. I was a teaching assistant during my first year of graduate school, and I found teaching to be the most rewarding and stimulating work I had ever done. I always loved helping others, and I always knew I wanted to teach to some degree, but that initial teaching experience made me want to spend the vast majority of my energy in teaching. I decided to go to DePaul University after receiving my Ph.D., but not as a postdoctoral researcher. I enrolled in their graduate teaching certification program and became certified to teach physics in high school.

I started my teaching career in the Chicago Public Schools, at an inner-city high school that had a ninety-five percent poverty rate amongst its students. I did not even apply to any other school district because I wanted to help those kids who needed the most help and support from caring adults. After three and a half years in Chicago, my family and I decided to move out to the suburbs and I got a teaching position at Evanston Township High School. My current position involves teaching advanced placement, calculus-based physics as well as mentoring and sponsoring students in independent science research for national competitions. I am also able to be a part of various collaborations with groups at Northwestern University. For me, going into high school science teaching has absolutely been the right career choice.

II. Description of Job Duties

It has been documented that a career in teaching is one of the most stressful professions. I believe the single biggest reason for this is the large number of small jobs that make up the complete job of a teacher. These small jobs may include terms such as counseling, advising, coaching, writing, sponsoring, chaperoning, disciplining, recommending, planning, mentoring, serving (on committees), consulting, researching, grading, pushing, prepping, modeling, attending, motivating, learning, sharing, and, in a bit of spare time, teaching.

Public high school science teachers normally have five classes per day and interact with over one hundred students per day. This can vary from school to school. Day-to-day responsibilities range from keeping track of attendance to planning, designing, and setting up daily lessons and laboratory experiments. Teachers are expected to keep students aware of their progress through a wide variety of assessments, and there are constant demands from students and their parents for as much individual attention as possible. In addition to their work in the classroom, many teachers spend large amounts of time before and after school sponsoring and coaching any number of clubs, activities, or teams. There are additional committee responsibilities as well as work within the department and staff development activities. Many teachers will take additional classes in the evenings or over the summer to stay current in their field. Typically, the more education a teacher has, the higher he or she can go on the pay scale.

As in any job, there is some amount of paperwork that needs to be done, but ideally the vast majority of time for a teacher is spent with students. The primary duty of a teacher is to get kids to think and learn, the rest of the many duties and responsibilities that come with teaching are secondary. Nothing is as rewarding as seeing a student's eyes light up when he or she finally understands something, and that is where the bulk of a teacher's energy is directed.

III. Advantages of Career

The advantages of becoming a science teacher begin and end with helping create scientifically literate and productive members of modern society. Teaching involves working with young people and getting them to think critically and analyzing the problem at hand. Our children will be the future, and I'll bet every person who is considering a career in teaching can remember a favorite teacher who helped turn them into the person they are now. Teachers can influence and change lives. A teacher can be the single greatest role model in a student's life, and there will be some number of children who become better people due to certain teachers. I personally cannot think of a more important job in our society.

Another major advantage and 'perk' of going into teaching is the summer. A school year is absolutely draining. It is difficult to take on the amount of responsibility that teachers take on. It is stressful working with over one hundred individual students each day, let alone trying to give individual attention to each of them. Teachers truly do need the summer to re-energize for another school year. But the summer is also a time many teachers can pursue other interests. This is a benefit not many other professions can offer or compete with. Teachers I have known have participated in science research with university groups, taught in summer school and other summer programs for additional money, worked with stock brokers and real estate agents, took additional college courses to pursue advanced degrees, or simply spent time relaxing or traveling with their families. One thing to avoid, however, is to go into teaching in order to have your summer off. This type of teacher will most likely end up being ineffective and could have a negative impact on students and colleagues alike.

IV. Essential Related Skills for Success

The most obvious skill one needs to teach science is a good understanding of the scientific subject matter in the curriculum. However, there are numerous other skills that are not as obvious that are necessary to be an effective teacher. I have found that these other skills include being able to think on your feet, developing a wide variety of different ways of presenting the topics being studied in the class, and forming relationships with adolescents that are built on mutual respect and trust.

Many teaching candidates as well as new teachers tend to believe that simply talking about new material will allow students to learn. What effective teachers learn through some number of years of experience is that there is no one way of teaching all the students in a particular class. Every single person learns best in his or her own unique way, and it is up to classroom teachers to recognize and accept this fact if they want to reach all their students. Over a period of time good teachers develop a repertoire of varied activities in order to present course material in ways that will benefit each individual learner. This requires a good deal of patience, but also good observation and evaluation skills. There are numerous books and educational research journals that discuss theories of learning and, one conclusion that appears over and over is this very idea that there seem to be stages of learning as well as different types of intelligence, and teachers need to provide a variety of activities to reach all students. Science teachers need to identify activities, methods, and instructional techniques that students are comfortable with through observing and evaluating what students and other colleagues are learning. For example, good, effective science teachers not only make use of 'traditional' lecture techniques, but they allow time for students to practice problem solving individually and in small groups, teacher- and student-centered demonstrations of the phenomenon being studied, class discussions and debates about the validity of certain conclusions from experiments and theories, making connections between the material being studied and everyday life, and, of course, laboratory experiments (after all, experimentation is what separates science from philosophy and religion). There are, of course, many other types of activities to do in class with students in order to provide learning opportunities for every individual student. Lastly, there needs to be a willingness to throw out ideas and teaching methods that do not work.

V. Additional Advice

When going into a teaching career, particularly teaching science, please take student teaching very seriously. This is where the teaching candidate gets to experience and practice what he or she will be preaching to students one day: one of the best ways to learn is by doing. Student teaching is where most teachers truly learn teaching skills. Perhaps the single most important aspect of student teaching is to develop good relationships with and learn from experienced teachers. It is from them that a new teacher will learn the "tricks of the trade" and gain invaluable insights into the nature of kids and how to approach the profession. Even if you do not get paid monetarily during your student teaching experience, you will find that you have been paid with knowledge and experience that will last your entire career.

Another bit of advice I can offer is that prospective teachers need to realize that learning is a lifetime activity both for their students and themselves, and that one can never stop trying to perfect their craft. Teaching is the most intellectually challenging activity I have ever taken on because of the enormous number of roles we must play during a single day as well as trying to figure out how to truly reach every student who attends the class. This requires patience, understanding, and a willingness to try something new for that student who just doesn't understand a concept after being exposed to the idea in a variety of ways.

The last advice I would give is that anyone who wants to go into science teaching should try to have at least minimal laboratory research experience. This will help solidify the entire subject trying to be taught. The teacher or teaching candidate will have firsthand knowledge of what the scientific method is and the way the process of science works. This knowledge and experience can, in my opinion, only make for a better teacher of science. If you are considering a career in science teaching, the single biggest reason should be because you are looking to help young people grow and become better thinkers, problem solvers, and productive members of a technologically based society. I cannot think of a more rewarding profession, where it is your job to help change lives for the better, and where a little piece of you branches off inside every student who passes through your classroom.

VI. Pitfalls to Avoid

As mentioned above in *Essential Related Skills for Success,* ideally those who set their sights on science teaching as a career are doing so primarily to help students. One must understand however, that states and school districts set standards and criteria that teaching candidates need to meet and satisfy. Many times the requirements for certification seem to serve more as hoops to jump through and consume vast amounts of time and money than useful experiences that will get you into a classroom so you can begin to actually work with students.

I have known several teaching candidates who became so fed up with waiting to complete coursework and get into a school to begin student teaching that they almost let their impatience win out and almost went into other careers. These teachers turned out to be very effective because the source of their impatience was the desire to take on one of society's single most important jobs. It is frustrating to go through state certification bureaucracy, courses, student teaching, background checks, testing, and the other activities that are required before actually getting that first job and first set of students in front of you. If you have the teaching bug and cannot wait to get in and hopefully change student's lives for the better, try to gather enough patience in order to fulfill those criteria. If you are passionate about helping young people in science, let the thought of the end result, which is you in a classroom with students, keep you going through the certification and the hiring processes so you can unleash your desire to help kids develop into mature, problem-solving, and critically thinking young adults.

VII. Qualifications Required

In order to teach in public high schools a prospective teacher will need at least a bachelor's degree in the field he or she will be teaching. In addition, states require other education courses and student teaching experience before being certified to teach in that particular state. States can vary the requirements necessary for teaching certification, and those who are considering a career in teaching need to check with the appropriate state board of education to learn more about the specific teaching requirements in the state you plan to teach. Most states are now at least considering 'fast-track' certification for teaching candidates from other professions due to the accelerating teacher shortage, especially in math and science.

ⓘnteresting Facts

❶ Many teaching employment opportunities will be available in the next ten years due to the number of teachers expected to retire.

❷ Information on science teaching can be acquired from: *http://www.nsta.org*.

❸ Although licensure is required by all fifty states, licensure may not be required for teachers in private schools.

Source: U.S. Department of Labor. *Occupational Outlook Handbook*, 2002–03 ed. Indianapolis, IN: JIST Publishing, 2002.

Take Home Points

- The primary duty of a science teacher is to get students to think and learn about classical and modern scientific principles
- The advantages of becoming a science teacher begin and end with helping create scientifically literate and productive members of society
- Every single person learns best in his or her own unique way, and it is up to classroom teachers to recognize and accept this fact if they want to reach all their students
- If you are considering a career in science teaching, the single biggest reason should be because you want to help students grow academically, become better thinkers, and better problem solvers
- Exercise patience when matriculating in the teacher-training program
- States can vary the requirements necessary for teaching certification; those considering a career in science teaching should check with the appropriate state board of education to learn more about the specific teaching requirements in the state you plan to teach

Suggested Reading List

Fried, R. *The Passionate Teacher*. Boston, MA: Beacon Press, 2001.

Glasser, W. *The Quality School Teacher*. New York: Harper Collins Publishers, 1998.

Sarason, S. *You Are Thinking of Teaching?* San Francisco, CA: Jossey-Bass Publishers, 1993.

VIROLOGY

> I maintain that cosmic religious feeling is the strongest and noblest incitement to scientific research.
>
> —*Albert Einstein*

Dr. Sherman S. Hom

West Nile Virus & Bioterrorism Lab Coordinator
Division of Public Health & Environmental Laboratories
New Jersey Department of Health & Senior Services

I. Biographical Sketch

I can still vividly see myself performing my first experiment at the age of six. After watching Mr. Wizard on television, I confirmed that hot air rises by measuring the temperature at the floor and ceiling of my kitchen. Several key mentors and events have helped me along my long and winding path to its present location of the public health arena. An enthusiastic middle school teacher awakened my general interest in biology. My high school physiology instructor taught me in great detail how the cellular processes of different organisms functioned. I knew I wanted to be a scientist when I took undergraduate courses in biochemistry, molecular biology, and cell biology at the University of California, San Diego. As a student worker in different labs earning my way through college, I performed experiments and isolated proteins from the cells of different organisms, such as enteric bacteria, human (liver and blood), hamster (connective tissue), and green algae. I simply loved working with living organisms and I loved the challenge of trying to understand how they functioned.

After earning my bachelor of science in biology, I matriculated at the University of California, Davis. For my Ph.D. in microbiology, I conducted genetic research on nitrogen fixation in *Rhizobium japonicum*, the bacteria that form a symbiotic relationship with the soybean plant that allows the storage of protein in its seed. I was the first to construct many mutants (10,000) of this organism by transposon-mediated mutagenesis. The goal was to increase the protein content in the soybean seed. I was a postdoctoral fellow for three years in the department of biology at the Johns Hopkins University in Baltimore, Maryland. During this period, I received training in molecular genetics, which focused on cloning and characterizing genes involved in both carbon and nitrogen fixation in *Rhizobium japonicum*. After my postdoctoral experience, I pursued a career as an industrial scientist. I obtained an excellent position as a senior research scientist with the Henkel Research Corporation (Santa Rosa, CA), a wholly owned subsidiary of an international Fortune 500 company. During my five-year tenure with Henkel as a molecular microbiologist, I cloned, characterized, and overproduced bacterial genes that produce extracellular enzymes for the fats and oils industry. These proteins included a lipase (fat-degrading enzyme) that was resistant to both high acidity & alkalinity (rare in nature) and also high temperatures. This clone produced my first patent. This patent was subsequently licensed by a large German pharmaceutical firm for resolution of racemic mixtures of ester linkages during drug synthesis. Next, I cloned a high temperature and alkaline resistant protease (protein-degrading enzyme) that was developed in a record two years and incorporated in their laundry detergent. Over a ten-year period, Henkel has produced sixty million pounds of this protease worth one billion dollars.

During my academic and corporate work, I was thus responsible for two patents and contributed scientific papers to *Journal of Bacteriology, Applied and Environmental Microbiology, Methods of Enzymology, Journal of General Microbiology,* and *Trends in Biochemical Sciences.* Since my undergraduate days, I have been a member of the American Society for Microbiology.

After ten years in academia and seven in corporate research, I became restless. I feel that the future of the Earth is really in the hands of the youth, so I began to teach. Over the next five-year period, I taught molecular genetics and general biology (majors and non-majors) at the community college, state college, and postgraduate levels in Saint Maarten, Los Angeles and New Jersey.

After about five years of teaching I longed to return to the research laboratory. Fortuitously, a position with the New Jersey Department of Health and Senior Services became available. This is where I have been ever since. Early on I served as the West Nile virus (WNV) lab coordinator overseeing our state's WNV surveillance program to protect the health of our citizens. Both the state of New Jersey and the Centers of Disease Control and Prevention (CDC) support my work. WNV is a disease that is transmitted by the bite of an infected mosquito and can cause mild or fatal illness such as encephalitis (i.e., inflammation of the brain) or meningitis (i.e., inflammation of the lining of the brain and spinal cord).

After 9/11/01, however, I was appointed the New Jersey bioterrorism laboratory coordinator. My current responsibility is to help prepare our Public Health Laboratory to provide rapid and effective laboratory services in response to biological terrorist attacks. Our labs also have responsibility for rapid analysis and response to other infectious disease outbreaks and other public health threats & emergencies. I feel a responsibility to construct a quality molecular diagnostics program for our state. In conclusion, from my current vantage point, I can clearly see that my diverse research and teaching experience in industry and academia has contributed to the unique repertoire of abilities in science and with people that have made it possible for me to contribute to the "post-9/11" modernization of the public health laboratories. I feel a deep sense of responsibility toward the future, and a sense of appreciation as I look back on my previous experiences in microbiology and virology.

II. Description of Job Duties

State public health laboratory coordinators are responsible for all aspects of a diagnostic testing program for the sole purpose of protecting the health of all the state's people.

Surveillance Testing Program

Diagnostic testing is the mainstay for all laboratory coordinators at a public health laboratory. The lab coordinator must utilize a multitude of managerial and scientific skills such as directing technical staff, validating robotic instrumentation, and maintaining a successful testing program. Specifically, public health laboratory coordinators write grant applications to procure funds, create yearly budget estimates, purchase capital equipment, hire personnel, train existing and new supervisors, validate existing diagnostic assays, develop and validate new diagnostic tests, and conduct extracurricular research. Knowledge of various scientific disciplines, such as bacterial physiology, biochemistry, molecular microbiology, cell biology, immunology, and virology is utilized to supervise the various programs in a state public health laboratory. The latest modern technology and instrumentation is used, because the lab coordinator is ultimately responsible for the most rapid, specific, sensitive, and accurate testing. Techniques include real time polymerase chain reaction assays, time-resolved flourometric immunoassays, single nucleotide polymorphism assays, direct and indirect immunofluorescense assays, cell culture, and serological tests. Sophisticated instrumentation includes robots for automated nucleic acid isolation, automated DNA sequence analysis, and biochemical characterization. All diagnostic work deals with pathogenic bacteria and viruses that must be tested under stringent CDC-mandated Biosafety Level-2 to Level-3+ safety conditions. Testing for microorganisms against which we have no antibiotics, such as the Ebola virus, must be done in a CDC-mandated Biosafety Level-4 facility. Only CDC and the DOD have Level-4 facilities.

My primary duty is to coordinate the West Nile virus (WNV) testing of humans, horses, crows, and mosquitoes. The primary purpose of the state's WNV surveillance program is to prevent transmission of this virus from mosquitoes to both humans and horses. More specifically, WNV is an arbovirus (spread by mosquitoes), which was first discovered in the late 1930s in Uganda in East Africa and has subsequently spread to most of Africa, western Asia, the Middle East, and Europe. A more famous arbovirus is the yellow fever virus. WNV first arrived in the New World in New York City in the summer of 1999. For the first three years after its appearance in the U.S., fourteen people died and 150 individuals became infected. By 2002, the virus has spread all the way to the United States West Coast and into Central America. From January through September 2002, over 300 people died out of over 3,000 infected individuals. An effective WNV testing program requires a very large test capacity, very high assay throughput per week, and a rapid test turn around time. After validating two robots, real time reverse transcriptase-PCR assays to detect WNV in mosquitoes has increased eightfold in our lab to 750 assays per week with a turn around time of forty-eight hours. This is a vast improvement over an older test that required a turn around time of ten days. The forty-eight hour turn around time allows mosquito control agency personnel to effectively identify and kill adult mosquitoes that carry the virus before additional transmission.

Assay development and validation

Lab coordinators also direct the modifications of existing assays to increase sensitivity and decrease costs. For example, the existing real time RT-PCR assay can detect less than one plaque-forming unit at a cost of only seven dollars per test. This is much lower than the $40 price tag on the older test. Recently, the WNV lab group has developed an assay to detect two arboviruses simultaneously in the same reaction well.

Research

Lab coordinators also direct and supervise research in public health areas.

My current research project is on the effect of various genetic polymorphisms in three human liver genes on the incidence of childhood leukemia.

III. Advantages of Career

The advantages of being a state public health lab coordinator include the opportunity to perform public service each day. One has the opportunity to work with all types of pathogenic bacteria and viruses that cause disease, such as food borne illnesses (*E. coli, Salmonella, Shigella, Campylobacter, Listeria*), tuberculosis (*Mycobacteria*), Lyme disease (*Legionella*), AIDS (HIV), hepatitis (hepatitis viruses A, B, & C), and the flu (flu virus Hong Kong strain). I am constantly appreciative of the opportunity to participate in improving the efficacy of ongoing surveillance testing programs, which prevent and diminish the severity of disease outbreaks. These various testing programs are the crux of epidemiologic investigations, which identify disease outbreaks, discover their cause, and provide solutions to prevent their spread.

IV. Essential Related Skills for Success

A thorough knowledge and practical research skills in microbiology, virology, and immunology is the foundation for success. In addition, a solid background in mathematics, biological statistics, and use of various computer software programs (word processing, spreadsheet, and database utilization) is also necessary. Furthermore, personal skills, such as the ability to multitask, prioritize, persevere, communicate with a diverse group of individuals, maximize effectiveness, think creatively and critically, troubleshoot problems, and work under extreme pressure (since lives are sometimes at stake) will facilitate a better chance in having a successful career. Ability to write various types of documents from internal memos to scientific papers, and to speak at various venues from scientific conferences to press conferences is also essential. Besides the typical science courses included in most biological science majors programs, I suggest taking additional electives, such as statistics, computer programming and software training courses, public speaking, and technical writing.

A very important course that has been most helpful in my present career in public health is immunology. Immunology is a broad science that deals with the study of the cell-mediated and humoral aspects of the immune system. Many serological diagnostic tests are based on the use of antibodies to detect other antibodies produced by a host's immune response to recent infection. In addition, direct immunofluorescense assays detect various whole bacterial cells (*Legionella*), time resolved immunofluorescense assays to detect whole cells and spores (*Bacillus anthracis,* the causative agent of anthrax), and indirect immunofluorescense assays to detect viruses (WNV & rabies viruses). An understanding of the basic science of antibody structure, function, production, and isolation is essential for a modern public health laboratory director.

V. Additional Advice

Before deciding on an undergraduate major in a particular scientific discipline, such as public health, I strongly suggest identifying, initiating contact, and then conducting a dialogue with a mentor in the field. This individual can share one's personal experience, knowledge, and even suggest additional contacts to broaden one's network of advice so that one can make an informed decision.

Many high school students and undergraduates dream of a particular career, but surprisingly discover a dislike for that career during their tenure. To eliminate such tragedies, I strongly recommend during one's undergraduate career that you take several independent research courses in different disciplines such as biochemistry, molecular genetics, and virology. In my own experience at UCSD, I took several independent research classes and discovered that I really loved doing biological research. I sincerely felt that I was fulfilling my boyhood dreams and fueling my desire to simply want to know how living organisms functioned. During these experiences I performed optimization experiments involving the isolation of red blood cell membranes and subsequently purification of proteins using gel electrophoresis.

As a member of the American Society for Microbiology I was able to attend the national conferences. These experiences really broadened my understanding of the various possible careers within the field of microbiology. I attended symposia, seminars, research talks, and poster presentations concerning a wide variety of topics. I found these annual events very encouraging and educational.

VI. Pitfalls to Avoid

One myth that is solidly established in the American psyche is that in the twenty-first century, like our parents, we will have only one career. A few years ago, I read a futuristic book entitled "Megatrends 2000," which flatly stated that one would have three or more distinct careers in this new century. They clearly advised that one should consciously and rationally plan for these career changes and not be forced into them by circumstances.

For example, I began my public health lab career as the West Nile virus lab coordinator. Recently, however, I have been appointed the state bioterrorism lab coordinator. I am responsible for the laboratory preparedness and response to a bioterrorism event in NJ. This program has thirteen staff in different areas, such as microbiology, molecular genetic, information technology, quality assurance & control, and training the state's clinical laboratories. In the future, my crystal ball shows me several possibilities, such as becoming the Director of Public Health Services but I am especially interested in applying molecular genetics approaches to diagnostics in many programs in the public health field.

VII. Qualifications Required

Although a Ph.D. in virology, microbiology or a related discipline is not required for many lab coordinator positions in some states, this postgraduate degree would enhance one's opportunities in obtaining very competitive lab coordinator positions in prestigious state public health laboratories, such as the ones in New York and Minnesota. Again, although not absolutely required, a postdoctoral fellowship, where one acquires additional knowledge and training in clinical microbiology, immunology, or virology would definitely be a plus. Preparation for a career in public health usually minimally requires a bachelor's degree in microbiology.

ⓘnteresting Facts

❶ Virology is a branch of science that deals with the chemical composition, structure, replication, and diseases associated with viruses.

❷ A career in virology will be promising due to the increase in viral based diseases around the world.

❸ More information about a career in public health can be obtained from: *http://www.asph.org*.

Source: U.S. Department of Labor. *Occupational Outlook Handbook,* 2002–03 ed. Indianapolis, IN: JIST Publishing, 2002.

Take Home Points

- State public health lab coordinators essentially directs a laboratory group responsible for diagnostic testing programs to detect disease-causing microorganisms or viruses in a human population
- State public health lab coordinators have a fulfilling opportunity as public servants to directly protect the health of a large population of people
- Organizational & managerial skills, multi-tasking, prioritizing, and time management skills are fundamental in gaining success in this field
- An open, honest, and trusting dialogue with a chosen scientific mentor is crucial in making career decisions
- Dialogue with individuals in the public health laboratory system who have had careers in the biological sciences or within the public health system can be very beneficial
- Obtaining the Ph.D. degree in microbiology, immunology, virology, or molecular biology is highly recommended for qualification as a laboratory coordinator

Suggested Reading List

Garrett, L. *Betrayal of Trust: The Collapse of Global Public Health.* New York, NY: Hyperion Press, 2001.

Koop, C., Pearson, C., & Schwarz, M. *Critical Issues in Global Health.* Hoboken, NJ: Jossey-Bass Publishers, 2000.

McCormick, J., Fisher-Hoch, S., & Horvitz, L. *Level 4: Virus Hunters of the CDC.* New York, NY: Barnes and Noble Books, 1999.

CHAPTER 4

Science Careers: Additional Resources

Now that you have completed this book and have begun to examine issues important to you as you pursue your science career let's now explore additional information that will allow you to derive maximum benefit from this book and allow you to have a healthy and productive career search. Specifically, this section contains the following items:

- Exploring the Internet
- Career Evaluation Form
- Science Career Network Investigation Form
- Science Career Network Questions
- Science Professional Associations
- Journal

The career exploration process can't be completed in one or two days; on the contrary, this is a process that is best accomplished over a period of months. The best advice regarding career exploration is to begin this interactive process as soon as possible. As I stated earlier, the end result of this momentous endeavor is too important to perform in haste. Lastly, it has been my pleasure assembling this career guide, and I wish you well as you embark on your career exploration journey.

Exploring the Internet

The Internet can make your career search easy and fun. The Internet can be an informative resource for researching prospective careers, conducting job searches, and posting resumes. By surfing the Internet you can find hundreds of helpful websites to assist you as you pursue the perfect science career. Most of the websites found in this section will contain additional links to other sites that will provide more information about a particular topic. While there are many websites about careers, I have selected only those I feel will maximize your use of the Internet and produce the results you want. Below are reliable websites grouped according to scientific discipline to help save you time as you access the information you need. Under each website address is a brief description of the type of information found in each site.

General Career Websites

CareerLab
http://www.careerlab.com/letters
This site contains many cover letters for different situations and occasions that are designed to meet all your needs.

Career Magazine
http://www.careermag.com
This site allows you to post your resume, conduct job searches, locate job fairs, and more.

Career Builder
http://www.careerbuilder.com
This site offers information designed to enhance your career search.

Career Planning
http://www.career-planning.org
This extremely helpful site offers a great selection of pertinent topics and useful links as you plan for your future career.

Chronicle of Higher Education Career Network
http://www.chronicle.com/jobs
This site is the premier place for those interested in a faculty, executive, or administrative position in the higher education system.

Free Resume Tips
http://www.free-resume-tips.com/10tips.html
This site offers ten free resume tips you can follow as you prepare the perfect resume.

Guide to your Career
http://www.guidetoyourcareer.com
This very informative site presents a useful career exploration approach and offers useful career links.

Job Web
http://www.jobweb.com/home.cfm
This site features helpful career development information and interesting career news.

Monster.com
http://www.monster.com
From this site you can post your resume, learn about new job opportunities, receive resume tips, and much, much more.

Occupational Outlook Handbook
http://www.bls.gov/oco
This site allows you to find important career information according to occupation.

ResumeEdge
http://www.resumeedge.com
This site contains a variety of sample resumes and cover letters. In addition, this site presents job openings and career links.

Salary.com
http://www.salary.com
This site offers many different options from examining job descriptions to performing salary projections in your career of interest.

Biology

BiologyJobs.com
http://www.biologyjobs.com
This site is designed for students or professionals looking for a job in the life sciences.

Careers in Biology
http://www.biodeveloper.com/~careers
This site presents information pertinent to those who want to learn about the requirements for entry into biology careers.

The Society for Integrative and Comparative Biology—Careers in Biology
http://www.sicb.org/careers
This site features a plethora of general career information about many different and exciting careers in biology.

Biotechnology

Bio-Jobs.com
http://www.bio-jobs.com

This site allows you to conduct job searches and presents tips on resume writing and interviewing.

BioView
http://www.bioview.com
This site offers information for persons interested in pursuing a career in the biotechnology and pharmaceutical industries.

SciWeb Biotechnology Career Center
http://www.biocareer.com
This site allows you to post your resume, conduct job searches, and offers excellent links.

Chemistry

ChemWeb.com
http://www.chemweb.com
This site provides useful information to chemists such as research tools and employment listings.

ChemStudent.com
http://www.chemstudent.com
This site allows students the opportunity to locate internships, jobs, and other research experiences in chemistry.

Chemjobs
http://www.chemjobs.net
This site features resume posting, worldwide job searching, and provides useful links.

Engineering

Engineering Central
http://www.engcen.com/jobbank.htm
This site presents a listing of jobs available in engineering as well as other useful resources for the engineering student.

EngineeringJobs.com
http://www.engineeringjobs.com
This site presents employment in engineering fields and provides useful links for engineers.

American Society for Engineering Education
http://www.asee.org
This site provides information about the American Society for Engineering Education.

Geology

American Geological Institute
http://www.agiweb.org

This site features information about the American Geological Institute and provides career and educational resources in the geosciences.

GeoScienceJobs.com
http://www.geosciencejobs.com
This valuable site presents employment opportunities for geoscience professionals.

Society of Exploration Geophysicists
http://www.seg.org
This site provides information about the Society of Exploration Geophysicists and offers excellent educational links.

Genetics

National Society of Genetic Counselors
http://www.nsgc.org
This site provides information about the National Society of Genetic Counselors and presents helpful career information.

The Human Genome Project Information
http://www.ornl.gov/hgmis
This site offers informative facts about the Human Genome Project and has valuable career and educational links.

American Society of Human Genetics
http://www.faseb.org/genetics/ashg/ashgmenu.htm
This site provides information about the American Society of Human Genetics and presents information a geneticist would find extremely useful.

Medicine

HealthCareSource.com
http://www.healthcaresource.com
This site features a wide array of job listings in the health care profession.

JobScience.com
http://www.jobscience.com
This site contains excellent employment information for those seeking a career in health care.

Association of American Medical Colleges
http://www.aamc.org
This site provides valuable educational resources for medical students.

Medical Technology & Microscopy

The American Microscopial Society
http://www.amicros.org
This site presents information on the American Microscopial Society and contains links for professional development for microscopists.

Medical Lab Careers
http://www.ascp.org/bor/medlab
This website contains information about the American Society for Clinical Pathology as well as information about laboratory careers, wages, and vacancy rates.

American Medical Technologists
http://www.amt1.com
This site contains information about the American Medical Technologists association and provides career information.

Microbiology

Careers in the Microbiological Sciences
http://www.asmusa.org/edusrc/edu
This is a very informative web page located in the American Society for Microbiology website.

Society for Industrial Microbiology
http://www.simhq.org
This site contains important information about the Society for Industrial Microbiology and also offers helpful information to the aspiring microbiologist.

The Microbiology Network
http://microbiol.org/jobs.HTM
This site provides a collection of links for those interested in a career in microbiology.

Molecular Biology

The Bio-Web
http://cellbiol.com
This site contains many educational and professional resources for the molecular biologist.

Molecular Biology Today
http://www.molbio.net
This site provides access to a scholarly journal that covers fascinating topics in molecular biology.

Molecular Biology Gateway
http://www.horizonpress.com/gateway
This site contains numerous links and educational resources for those interested in molecular biology and other sciences.

Nursing

4 Nursing Jobs.com
http://4nursingjobs.com
This site contains information for those interested in a career in nursing and other health professions.

National League for Nursing
http://www.nln.org
This site provides information on the National League for Nursing as well as important career links for potential nurses.

All Nursing Schools
http://www.allnursingschools.com
This helpful site contains information about nursing colleges in the U.S.

Pharmacy

PharmWeb
http://www.pharmweb.net
This site contains a directory of pharmacy schools in the U.S. as well as numerous educational links to health care and pharmaceutical professions.

RXinsider.com
http://rxinsider.com
This valuable site contains an extensive listing of current pharmacy jobs in the U.S.

Pharmacy Choice
http://www.pharmacychoice.com
This site offers career information and many educational resources for students interested in pharmacy.

Physics

PhysicsWeb
http://physicsweb.org
This site presents current employment opportunities in the physical sciences and offers additional educational resources.

PhysLink.com
http://www.physlink.com
This site contains employment information and other pertinent information for physicists.

GradschoolShopper.com
http://www.gradschoolshopper.com
This site contains informative information for those students interested in pursuing a graduate degree in physics.

Virology

Institute for Molecular Virology
http://virology.wisc.edu/IMV
This site offers educational and professional resources for students interested in virology.

American Society for Virology
http://www.mcw.edu/asv

This site contains information about the American Society for Virology and provides useful career resources in virology.

All the Virology on the WWW
http://www.virology.net/garryfavwebindex.html
This site presents many links to learn about the exciting field of virology.

Other Science Websites

Science Careers
http://recruit.sciencemag.org
This helpful site contains current employment information in the sciences and offers employer profiles, resume posting, and more.

Science Jobs
http://www.sciencejobs.com
This site contains information for those interested in either bioscience- or chemistry-related fields.

Graduate Schools for the Sciences
http://www.gradschools.com/medicalsearch.html
This site helps you locate information on the graduate schools that offer your specific graduate program in the sciences.

Science—Next Wave
http://nextwave.sciencemag.org
This helpful site provides career resources for scientists.

Science Careers Web
http://www.sciencecareersweb.net
This useful site provides helpful career resources for students interested in a career in the sciences.

Career Evaluation Form

____

Science Career

Advantages of Career	Disadvantages of Career
1.	1.
2.	2.
3.	3.
4.	4.
5.	5.
6.	6.
7.	7.
8.	8.
9.	9.
10.	10.

Educational & Personal Goals

1 ______________________________

2 ______________________________

3 ______________________________

4 ______________________________

5 ______________________________

Science Career Network Investigation Form

Prospective Science Career: ____________________________

Contact Information

Name:

Address:

Telephone Number:

Fax Number:

Job Title of Contact, Institutional Affiliation:

E-mail Address:

Additional Notes:

____

____

____

____

____

Science Career Network Questions

The following is a list of questions you can pose to members of your career network. Use these questions when conducting interviews, when writing letters, or sending e-mails. As this list is by no means exhaustive, use these questions as a guide to create your own personalized questions for career exploration.

1. What are the advantages of working in your particular career?
2. What are the disadvantages of working in your particular career?
3. What other courses (science, non-science) or related experiences should I consider in order to acquire the skills I need to be successful in your career?
4. Are you aware of internship possibilities, job openings, or any other opportunities for training?
5. What pitfalls associated with your career would you advise me to avoid?
6. Are you aware of any other persons that may be able to assist me in pursuit of information concerning my prospective career?
7. Are you aware of any books, magazines, articles, or journals I would find useful in acquiring more information about my prospective career?

8. Can you give me a general description of your actual (daily) job duties?
9. Can you provide me with information regarding the educational (e.g., degrees) and professional requirements for entry and success in your field?
10. Are you aware of any scholarships, fellowships, or grants available to students majoring in your field?
11. What additional insights can you offer that may prove helpful to me as I consider a science career. Specifically, how have your professional development experiences (e.g., educational, work, etc.) helped you prepare for your career?
12. What undergraduate, graduate, or professional schools will offer me the best education and overall learning experiences as I prepare for my science career?
13. Are you aware of any volunteering or shadowing opportunities that will allow me to learn more about your science career?
14. What helpful methods or approaches can you offer as I prepare for my entry exam (e.g., college, company)?
15. Can you recommend any videos, multimedia software, workshops, or professional meetings that will allow me the opportunity to learn more about your current career?

16. Specifically, what experiences shaped your decision to pursue your current career?
17. If your current career is not the career you initially chose, what events lead to your decision to pursue a different occupation?
18. What do you most enjoy about your current science career?
19. What do you least enjoy about your current science career?
20. In retrospect which of your many college courses were most helpful in terms of preparing for your current science career?
21. What helpful tips could you offer to assist me in preparing for career fairs?
22. Are you a member of any professional associations? If so, which ones? How have these professional associations positively affected your professional development?
23. What are some possible obstacles, experiences, or barriers I may encounter (during my educational career) as I pursue your current career? Could you offer positive solutions I can use to overcome them?
24. If you could start your educational or professional career over again, what would you do differently (e.g., high school, undergraduate, graduate school) that would better prepare you for your current career?

Science Professional Associations

Below are addresses of a number of major national science professional associations for you to explore. Contact the professional association of your choice to obtain information about your prospective science career. Also, contact the associations for information regarding membership, internships, research experience, publications, and other educational services.

American Chemical Society
1155 Sixteenth Street, NW
Washington, DC 20036
(800) 227-5558
E-mail: help@acs.org
http://www.chemistry.org/portal/Chemistry

American Dental Association
211 E. Chicago Avenue
Chicago, IL 60611
(312) 440-2500
http://www.ada.org

American Dietetic Association
120 South Riverside Plaza, Suite 2000
Chicago, IL 60606-6995
(312) 899-0040
E-mail: membrshp@eatright.org
http://www.eatright.org

American Institute of Biological Sciences
1444 Eye Street, NW, Suite 200
Washington, DC 20005
(202) 628-1500
E-mail: admin@aibs.org
http://www.aibs.org

American Institute of Chemical Engineers
3 Park Avenue
New York, NY 10016-5991
(212) 591-7338
E-mail: xpress@aiche.org
http://www.aiche.org

American Institute of Medical and Biological Engineering
1901 Pennsylvania Avenue, NW, Suite 401
Washington, DC 20006
(202) 496-9660
E-mail: info@aimbe.org
http://www.aimbe.org

American Institute of Physics
One Physics Ellipse
College Park, MD 20740-3843
(301) 209-3100
E-mail: aipinfo@aip.org
http://www.aip.org

American Institute of Professional Geologists
8703 Yates Drive, Suite 200
Westminster, CO 80031-3681
(303) 412-6205
E-mail: aipg@aipg.org
http://65.218.13.180/ScriptContent/Index.cfm

American Medical Association
515 N. State Street
Chicago, IL 60610
(312) 464-5000
http://www.ama-assn.org

American Nurses Association
600 Maryland Avenue, SW, Suite 100
Washington, DC 20024
(800) 274-4262
E-mail: memberinfo@ana.org
http://www.nursingworld.org

American Pharmaceutical Association
2215 Constitution Avenue, NW
Washington, DC 20037-2985
(202) 628-4410
E-mail: membership@mail.aphanet.org
http://www.aphanet.org

American Physical Society
One Physics Ellipse
College Park, MD 20740-3844
(301) 209-3200
E-mail: membership@aps.org
http://www.aps.org

American Society for Biochemistry and Molecular Biology
9650 Rockville Pike
Bethesda, MD 20814-3996
(301) 634-7145
E-mail: asbmb@asbmb.faseb.org
http://www.faseb.org/asbmb

American Society of Civil Engineers
1801 Alexander Bell Drive
Reston, VA 20191-4400
(800) 548-2723
E-mail: cybrarian@asce.org
http://www.asce.org

The American Society for Clinical Laboratory Science
6701 Democracy Boulevard, Suite 300
Bethesda, MD 20817
(301) 657-2768
E-mail: ascls@ascls.org
http://www.ascls.org

American Society of Mechanical Engineers
Three Park Avenue
New York, NY 10016-5990
(800) 843-2763
E-mail: infocentral@asme.org
http://www.asme.org

Association of Clinical Research Professionals
500 Montgomery Street, Suite 800
Alexandria, VA 22314
(703) 254-8100
E-mail: office@acrpnet.org
http://www.acrpnet.org

Biotechnology Industry Organization
1225 Eye Street, NW, Suite 400
Washington, DC 20005
(202) 962-9200
http://www.bio.org

Genetics Society of America
9650 Rockville Pike
Bethesda, MD 20814-3998
(866) 486-4363
E-mail: babbott@genetics-gsa.org
http://www.genetics-gsa.org

Geological Society of America
P.O. Box 9140
Boulder, CO 80301-9140
(888) 443-4472
E-mail: membership@geosociety.org
http://www.geosociety.org

Institute of Industrial Engineers
3577 Parkway Lane, Suite 200
Norcross, GA 30092
(800) 494-0460
E-mail: cs@iienet.org
http://www.iienet.org

Microscopy Society of America
230 East Ohio, Suite 400
Chicago, IL 60611
(800) 538-3672
E-mail: BusinessOffice@msa.microscopy.com
http://www.microscopy.com

National Science Teachers Association
1840 Wilson Boulevard
Arlington, VA 22201-3000
(703) 243-7100
http://www.nsta.org

National Association of Biology Teachers
12030 Sunrise Valley Drive, Suite 110
Reston, VA 20191
(800) 406-0775
E-mail: office@nabt.org
http://www.nabt.org

The American Society for Cell Biology
8120 Woodmont Avenue, Suite 750
Bethesda, MD 20814-2762
(301) 347- 9300
E-mail: ascbinfo@ascb.org
http://www.ascb.org

The American Society for Microbiology
1752 N Street NW
Washington, DC 20036
(202) 737-3600
E-mail: Membership@asmusa.org
http://www.asmusa.org

Journal

Use the following pages to record any pertinent information during the career exploration process.

INDEX

ABOUT THE EDITOR

Lawrence O. Flowers is a doctoral candidate in the department of microbiology and cell science at the University of Florida in Gainesville, Florida. He received his bachelor of science in biology from Virginia Commonwealth University in Richmond, Virginia and a master of science in science education and a master of science in biological sciences from the University of Iowa in Iowa City, Iowa.

A popular teacher, he is well respected by his students for his innovative pedagogical approach to teaching science, receiving an outstanding graduate teaching award from the University of Iowa during his graduate tenure. Mr. Flowers is also a member of several national professional organizations including the National Science Teachers Association and the American Society for Microbiology.